The Emergence of Video Processing Tools

Television Becoming Unglued

Volume 2

The Emergence of Video Processing Tools
Television Becoming Unglued

Volume 2

edited by Kathy High, Sherry Miller Hocking and Mona Jimenez

intellect Bristol, UK / Chicago, USA

First published in the UK in 2014 by
Intellect, The Mill, Parnall Road, Fishponds, Bristol, BS16 3JG, UK

First published in the USA in 2014 by
Intellect, The University of Chicago Press, 1427 E. 60th Street,
Chicago, IL 60637, USA

A catalogue record for this book is available from the
British Library.

Cover image: Photo courtesy Experimental Television Center Archives.

Production manager: Tim Mitchell
Cover designer: Ellen Thomas
Copy editor: Michael Eckhardt
Typesetting: John Teehan

ISBN 978-1-84150-663-0
ePUB ISBN 978-1-78320-301-7
ePDF ISBN 978-1-78320-300-0

Printed and bound by Hobbs the Printers Ltd, UK

Dedication

For those who have passed on – Barb Abramo, Connie Coleman, Evangelos Dousmanis, Dara Greenwald, Bill Hearn, David Loxton, Don McArthur, Phil Morton, Nam June Paik, Mary Ross, Steve Rutt, George Stoney and Jud Yalkut. Your work helped shape this project and still inspires.

And for the artists and technologists who have contributed to the creation of instruments and those who continue to do so

Contents

SECTION 3: TOOLS

SECTION 3

TOOLS

Introduction

Mona Jimenez

In its broadest definition, a tool is defined as 'something used in the performance of an operation; an instrument' (Morris 2000). In the 'Tools' section, the texts speak to the devices and operations used by artists to produce and alter video and audio: the nature and functions of the machines and processes, how they were designed and shared, and how the tools might remain 'alive' and functional. The reader may wish to reference the Experimental Television Center's web site 'Video History Project – Tools' for additional information, such as tool descriptions, diagrams, manuals, and photographs about the specific tools mentioned in the essays in this section.[1]

The section opens with Kathy High's 'Mods, Pods and Designs: Designing Tools and Systems', which lays out the impulses and motivations to hack, modify and invent custom video tools and describes the various relationships among artists, tool designers, engineers and enthusiasts that resulted in such tools as the Paik/Abe Video Synthesizer, the Rutt/Etra Video Synthesizer, the Jones Buffer, and software packages Image/ine and Jitter.

In the 1970s and 1980s, the Experimental Television Center (ETC) sponsored many grant proposals for software and hardware development; one of the most developed concepts (although never fully realized) was the Computer-Based Video Synthesizer System, carried out by Richard Brewster, Donald McArthur and Walter Wright; the final report is included here. The Computer-Based Video Synthesizer System was one of a host of projects, proposals and philosophies shared at a remarkable cross-disciplinary conference 'Design/Electronic Arts', held in 1977 in Buffalo, NY, and described here in an essay of the same name by its organizer, John Minkowsky. Organized under the auspices of Media Study/Buffalo, a nonprofit media arts center in Buffalo, NY, the conference brought together theoreticians, tool and system designers, and artists/users from the media arts, computer graphics, animation and modeling, media communications, neuroscience, electronic music and information science. Attendees included Gene Youngblood, Ken Knowlton, Laurie Spiegel, Joel Chadabe, Charles Csuri, Sonia Sheridan, Tom DeFanti, Tom DeWitt, Robert Ashley, John Whitney and Stan Vanderbeek.

In his essay 'Instruments, Apparel, Apparatus: An Essay in Definitions', Jean Gagnon places artist use of technological devices for sound- and image-making firmly within the realm of performance. Gagnon analyzes historical terminology used in material culture, music, and science to examine the way artists use electronic instruments to instantaneously enact visual and aural changes in real time. Also on the theme of definitions, Jeremy Culler's 'Expanding "Image-processed Video" as Art: Subverting and Building Control Systems' traces sources for the phrase 'image-processed video' that was formulated in the context of a critique of conventional commercial and educational television and quickly evolving categorizations of video art.

Sherry Miller Hocking's 'The Grammar of Electronic Image Processing' interweaves the vocabulary of video processing (as it appeared in various ETC technical publications), the words of artists and inventors, and descriptions of custom video tools and their development. Explanations of concepts such as signal, sync, switching, sequencing and keying found in ETC's manuals not only enabled artists to increase their technical understanding and thus navigate the open-ended ETC studio, but also documented the basic theories and language of image processing.

The theme of performativity is returned to in 'ETC's System' by Hank Rudolph, an artist and educator who worked for many years as the studio manager for ETC. Rudolph introduces the capability of ETC's studio, setting forth what he believes are its six fundamental characteristics that exemplify the system: the ability to affect image and sound in real time, an open-ended architecture, indeterminacy, interactivity, sound-image synchronization, and the ability to electronically generate images – a camera-less method of working. In the interview 'On Voltage Control' with Kathy High and Mona Jimenez, Rudolph further explores one of the core principles of video image processing – voltage control – that, like many design principles for video tools, was adopted from audio synthesis. Carolyn Tennant's 'Insofar as the rose can remember…' captures the feel of a bittersweet artist residency occurring just before the ETC studio closed, where she and her partner, J. T. Rinker, conjured and combined images of some of the key presences in ETC's history.

Dictionary definitions of tools still distinguish analog manifestations ('a device […] used to perform or facilitate manual or mechanical work') from digital (a software application that 'creates, manipulates, modifies, or analyzes') (Morris 2000). While one may expect a general trend in tool development where analog tools were eventually replaced by digital, contributors describe the parallel development and integration of digital and analog tools in the 1970s and 1980s. This integration was the norm at places such as ETC and the Electronic Visualization Lab at the University of Illinois at Chicago Circle Campus, where the Sandin Image Processor was used in combination with Z-GRASS, a computer graphics program developed by Tom DeFanti.

Yvonne Spielmann, in 'Analog to Digital: Artists Using Technology', reiterates a relationship between analog and digital processes in custom video tools, citing programmability, modularity, logic circuits and clock functions as examples of shared attributes. Spielmann associates specific functions of video tools with artists' works, particularly works by Steina and Woody Vasulka, Gary Hill and Peter Donebauer, and discusses parallels with earlier collaborations among artists and engineers through the group Experiments in Art and Technology (EAT). In the interview with designer Dave Jones in 'Analog Meets Digital In and Around the Experimental Television Center', and in articles on HARPO (software for voltage control) and Pantomation (an early motion-tracking device), the melding of the manual with the algorithmic is revealed with early custom video tools: the use of hand-turned knobs and patch programming was often combined with automated controls that created, manipulated and modified images through externally applied digital and analog signals.

Jimenez returns to the notion of tools as instruments in 'Preserving Machines', considering theories and practices of conservation as they apply to machines. Instrument conservation, the conservation of media-art installations and industrial conservation all deal with how the full function, design and purpose of a tool can remain visible and evident. The essay documents the efforts of advocates and experimenters who are working to keep tools functioning and available, such as through the 'Rutt/Etra Restoration Party', a fixers' event. 'A Catalog Record for the Raster Manipulation Unit' is next, an example of a cataloguing template that Jimenez designed as one way to document custom tools.

In 'Copying-It-Right: Archiving the Media Art of Phil Morton', Jon Cates gives voice to Morton's philosophies on the circulation of culture and ideas, reminding us to be free and generous with what we make and have learned. He discusses various forms of resistance to laws and policies that would limit the free exchange of ideas and creative work. His essay is paired with Ralph Hocking's manifesto on videotape preservation using a 'Resurrection Bus' with its emphasis on grassroots action.

References

Morris, William (ed.) (2000), *American Heritage Dictionary of the English Language*, 4th edition, Boston: American Heritage and Houghton Mifflin.

Notes

1. See Video History Project at http://www.experimentaltvcenter.org/history.

Mods, Pods and Designs: Designing Tools and Systems

Kathy High

> Creativity chooses to show up differently according to professional occupation: entrepreneur, artist, researcher or engineer. (Paik 2011a: 21)

How did artist-tool production get started in the United States? What is the relationship between artists, technologists and designers, and how can we describe the slippery ways these definitions of artist and technologist switch back and forth in the creation and development of new tools? Tool development and design is an ongoing and always-evolving process, with the technical needs being constantly redefined and the limitations or possibilities of the technology being explored. Successful collaborations between technologists and artists inspire each participant to have an investment in the contribution of the other. Composer and musician Philippe Manoury, who collaborated with software designer Miller Puckette, described an ideal working relationship: 'I think it's very important for a composer to have in front of him somebody [who] has also his own aesthetic idea. It sounds good or not good, and how we can ameliorate that, and to make suggestions' (Manoury et al. 2008). Thus each party from the artistic side and the engineering side contributes to the process of building something entirely new. This chapter looks at examples from artists and engineers/designers to think about the assorted methodologies used to build custom tools. To capture a feeling for how these processes evolve, I felt it was important to leave extensive quotations from artists and designers, allowing them to speak for themselves. Thus artists and designers speak in their own voices.

We have to ask: In what ways do artists and programmers think differently about creating software? How are their approaches similar? Tom Demeyer, creator of video processing software Image/ine explores this idea:

> Really fundamentally? No difference other than that between different artists. [...] Obviously programmers tend to think from the perspective of technical possibilities, artists from artistic potential. This ends up, often, in the artists having to restrain the programmer – at least, the good ones; others want more and more, thinking it will give them freedom of expression where in truth it becomes harder and harder to make good use of the tools. (Demeyer 2010)

Let us consider the history of this medium of video. Video, in the form of television, has been around for a while. Electronic television was released by RCA in 1939 at the World's

Fair in New York City, leaving its older cousin, the less resolved mechanical television, behind in the dust.[1] This was good for big corporations such as RCA, who took over the business of developing the technologies; bad for the tinkerers, hackers and developers who had been experimenting with mechanical TV transmitters and receivers. Since the late 1930s, electronic television dominated the communication waves, was maintained and developed only by a few companies, and thus was more difficult to build and alter.

In the United States by the late 1960s, television production equipment became smaller, portable and affordable, and thus began to get into a broader public's hands. Resourcefully, hackers and artists used scrapped materials or jerry-rigged found equipment to create new machines. By 1970 and on, there was pioneering video work being made – creating machines for art making. There were handfuls of teams across the country building, modifying and designing tools:

> In the long-ranged future, such a versatile color synthesizer will become a standard equipment like today's Hammond organ, or Moog's synthesizer in the musical field, but even in the immediate future it will find wide application. [...] [I]t might end up by producing a new fertile genre, called 'electronic opera'. (Paik 2011b: 9–10)

The 1960s and 1970s were particularly fruitful in contemporary art for art and technology explorations. It was also a period of extreme social unrest in the United States, with growing 'anti-establishment' sentiments. People were drawn to new models of labor, working in collectives, sharing resources and establishing a shared 'commons'. Artists and tool designers were interested in repurposing and scavenging, using scrap waste from industry to rebuild their own machines, and to experiment with those new machines to create original images. Working artists and technologists were crafting a vision of a creative culture, a do-it-yourself radical culture that opened up new ways of seeing the video medium. Artist Peer Bode describes watching the image output from a Jones Buffer at the Experimental Television Center:

> What a weird fucking image that is. I mean, that thing was video [...]. It had cinema, video, and then a keyer. It had aspects of all of those. That's what it was, right in the circuitry. But it made an image that none of us had seen before. [...] And so it created this wonderful problem for years, because it was a solution to a problem nobody had, you know? Sort of like, what are we going to do with this? (Bode 2012)

Peer Bode imagined the machine hacking of that period also as a form of tweaking culture and rethinking people's relationship to big business and to mass production. Notions of disturbance played into this art and technology development: much like the historical art movements of the Dadaists, Surrealists and Fluxus, artists became interested in the question of how to disrupt and provoke the status quo to encourage people to think

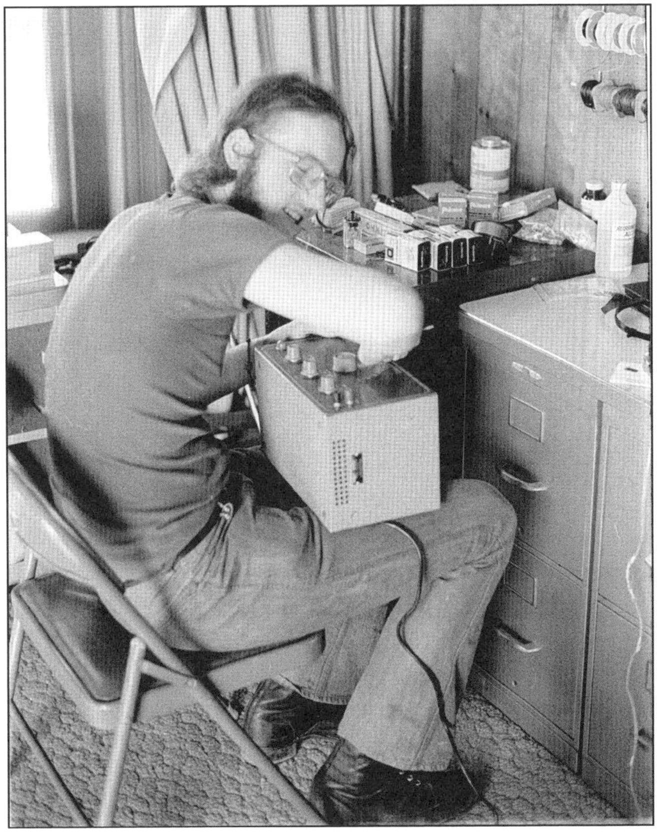

Figure 1. Dave Jones at workbench, Experimental Television Center (c. 1974).

critically. Similarly, gestures such as modifying and adapting industry tools became ways to interrupt the mainstream media message and create a new kind of signal.

Bode, who served as Artists' Coordinator at the Experimental Television Center (ETC) from 1978–87, was intrigued by the idea of 'hacking into things that already exist', and credited Ralph Hocking and Dave Jones with these 'adjustments' to machines, such as:

> [D]oing modifications of cameras, they were [...] adjusting how the Videcon tube responded, contrast, brightness, just delay, lag. So stuff that was on the board, that you would use a screwdriver for, Ralph had David made those things external, so that you could control them. (Bode 2012)

Bode points out the advantage to this kind of externalization of controls: 'It had extra controls that were brought on the outside, so again, you could get at some of the stuff that was in the circuitry. So this was this kind of hacking idea.'

Bode aptly describes the significance of this kind of modification of machines. It was a way to insert various cultural meanings and rereadings of the corporate 'norm' or 'mass media':

> I think, conceptually, it was also about the fact that there was some sense that television was technology, but it was also like a cultural form. And in a way, it meant the culture. You know, sort of like Nam June destroying violins wasn't just that he had a thing about violins, but it was about Western culture [chuckles] and Western history that he was making a gesture about. And I think that some of the kind of modifications of stuff were like that, in terms of the technologies. (Bode 2012)

One of the most important aspects of building machines was building the 'commons' and thus an information exchange that grew out of each network of artists and builders. People began trading tips and amateur tweaks that could be made to industrial machines. Out of these exchanges came documentation of systems, designs and even scores for creating new instruments.

At this time, New York State was a 'haven' for artist support with extensive state-based public and foundation funding, as well as institutional support, such as through university centers. And upstate New York (the area outside the New York City region) became a hot spot for innovation. In the 1970s, designer Dave Jones worked for a number of years with the Experimental Television Center, developing its tool set while aiding others to build tools as well. Jones spoke about the fluid ways that people began to find each other to gather technical advice:

> Back in those days, the people that were doing video, for the most part, had to at least learn a certain amount of tech to be able to do it, because there weren't readily available technicians to help out. It was new to pretty much everybody. And so just to be able to hook everything up, you had to do your own tech work. And when your equipment broke down, you either had to take it to someplace like Sony and get it fixed, or you had to talk to whomever you could talk to about getting materials and trying to fix it yourself. [...] Mostly, it was about networking. (Jones 2007)

Sources for sharing information in New York included the Raindance Foundation's *Radical Software* magazine (started in 1970) with technical tips and information on low-cost portable video equipment. Another source was a monthly newsletter called *Electronotes: Newsletter of the Musical Engineering Group*, published out of Ithaca by Bernie Hutchins. There were also electronics companies, such as Pia Electronics, offering DIY audio synthesizer kits that were used by many musicians, and PD Surplus, an electronics surplus store, among others. What appears to be an organic and innate process of DIY teaching and information exchange became what Jones calls '[a] fairly even flow of information [...] technicians would talk to other technicians about what they'd done and different ideas they had and, you know, it was a fairly even flow of information' (Jones 2007). These 'tech talk' exchanges reflected moments of community support.

The best collaborations often grew out of specific projects. Software designer Miller Puckette and musician/composer Philippe Manoury seemed to find a very comfortable middle ground between them where they both articulated their theoretical and the practical needs to each other. Manoury describes a keen balance that is necessary for artists and technologists to collaborate:

> [T]o make good research, it's necessary to have a musical project. [...] The program had a lot of scientists, scientific people who were working on computer music but were too proud of their systems or ideas. And it's necessary to have a theoretical point of view, it's absolutely important, but you also need a practical approach. We must have them both. (Manoury 2008)[2]

People were experimenting and creating tools that had not been built before. What was remarkable about this interdisciplinary approach from the 1970s onwards was that both artists and designers were pushing the video medium forward and discovering innovative ways to work with it. Technician/artist collaborations often involved volleying processes with much back and forth, give and take. Steve Rutt, engineer, technologist behind the Rutt/Etra Video Synthesizer, describes his relationship with his artist-designer collaborator Bill Etra:

> Bill would come in and he would go, 'I want to build this thing!' And I'd go, 'what does it do?' And he'd explain something – I might or might not get it. I mean, I would usually get what it was supposed to do, but I wasn't sure that the world needed it. And then I'd go back to trying to clean up the signal or make a better something on it to make it a little easier to build, or maintain, or to move, and Bill would like, stay on my case – 'come on, we need this something or other!' And eventually he'd badger me into, well, go ahead and prototype something. Back then we used to do everything in-house – we etched our own boards, we laid them out ourselves... pieces of tape – you didn't have a computer to do it back then. We would eventually have a new module that did something, and then we would run off with it and come back after a while with some product that was done and go, 'Look! See? See, it's cool!' (Rutt and Etra 2007)

The cool factor went far. People were into new territory and did not know what to expect from these machines. Peer Bode describes his reactions to the Paik/Abe Video Synthesizer at the Experimental Television Center, and how everyone had to grapple with this new kind of image making. He bemoaned the fact that he had not made more recordings with the device:

> The Paik/Abe [Synthesizer] was, like [...] nobody could figure it out. That was such an uncontrollable device. But it made great images. And so if you could handle just having a little bit of control, and also having – [...] Well, I definitely came out of a cinema kind of background, and with a certain sort of taste values around some of that stuff. And I know some of the first things I saw were with the Paik/Abe, and I couldn't believe how horrible I thought they were. [chuckles] They were just totally like, 'Oh, my God, what is this shit?' Which I think was the reaction of a lot of people. You know, because it was a kind of totally non-culture – no one had seen it. Again, now I feel embarrassed about my own kind of conservatism in my response. (Bode 2012)

Korean-born artist Nam June Paik was one of the pioneer tool builders in the United States and the Paik/Abe Video Synthesizer was his first video-processing machine. Paik had very concrete ideas of how he wanted his video synthesizer to work:

> Here are a few aspects of my conception: effective variations of scanning, in circles, in zigzag, complex feedback, mixing of audio signal, rescanning with ten black and white Sony video cameras, applying optic factor. And above all, I had conceived things in a way to create an open system with multi-input and multi-output, that is, an unclosed electronic environment that allows for public participation. It was an open system something like an archetype of an interactive video game. (Paik 2011a: 18)

Nam June Paik described the Paik/Abe Video Synthesizer as a 'sloppy machine, like me' (Park and Lee 2011: 38); 'sloppy' because it was unusual and unruly for the time, and it manipulated the video image in almost abstract ways. But far beyond its ability to produce unusual imagery, it was also an 'unclosed electronic environment that allow[ed] for public participation'. This philosophy of public participation was the basis of Nam June's artwork. He was creating a 'commons', and with this synthesizer device, a means for a broad public to interact with television, at that time of a closed one-way delivery system. He wanted to enable people to build a 'versatile color synthesizer' that would become 'standard equipment' (Paik 2011b: 9–10) in peoples' homes, so that they could control their own means of production – furthering participating on communication and creating a dynamic communication platform.

Paik collaborated with Japanese engineer Shuya Abe to build the Paik/Abe Video Synthesizer (PAVS). Paik met Abe in 1964 and they become long-standing collaborators after that. Paik found a rare collaborator in Abe. Abe spoke of his commitment to their work:

> I had a vague idea when Paik roughly described [the Paik/Abe Video Synthesizer], but it was a complete blank to me what a synthesizer was until September 1969. [...] Paik showed me a sheet of blueprint at home. When I saw it, I told myself 'Oh, I guess will have to quit my job in order to make this happen.' (Park and Lee 2011: 23)

In 1970, the Paik/Abe Video Synthesizer, which was built with funds from the WGBH New Television Workshop, was used in the WGBH studio for a four-hour live television broadcast entitled *Beatles from Beginning to End*. Abe helped build the synthesizer with specifications allowing it to stay within FCC norms and television broadcast regulation standards. In fact, Abe himself 'performed' the Paik/Abe Video Synthesizer for the entirety of the hugely successful live TV production. Nam June summed up the contradictory aspects of the experience by saying, 'Ironically a huge Machine [WGBH, Boston] helped me to create my anti-machine' (Paik 2011c: 11).

Paik and Abe's 'anti-machine' was among those at the forefront of artistic machine development at this time. Working closely with the development of the machines allowed artists to create works exploring how to disrupt the electronic signal itself. Messing with the signal was very much a similar process to modifying and adapting machines, building circuit boards and such. It was also a way to break down the system of authority, a corporate/capitalist system that dominated mass communication networks. Artists were literally 'breaking' the video signal, messing with the electronics in ways to distance their work as far from the monolith of television as possible.

In the evolution of analog video image–processing tools, one machine led to the development of the next version. There seemed to be steps taken by artists and engineers that promoted understanding and growth in building machines, as well as many failures between the design and building of a tool, and the ways to imagine it working. Steve Rutt talked about this one-step-forward-two-steps-backwards process: 'When we looked at what we were building [...] we kind of took a big step back before we even got to repeatability, and we said, "What are we trying to do here? We're trying to move an image"' (Rutt and Etra 2007). There are whole lineages of machines, a constant stream of updates and tweaks and new systems of ideas implemented. As artist-designer Matt Schlanger said: '[T]hen a lot of the gear were signposts, they were stepping stones along a process [...] manifestations of a process, manifestations along a process that were interesting at the time, but have no place in the studio now' (Schlanger 2012).

Certain machines were early versions of other machines; ones developed later better articulated the designer's ideas. How did designers think about versioning? Were some modifications and some finished designs? And in the future what machines will be preserved and revisited historically? Are they the final iteration of the final tool? Is that part of the process and design of interest? Did all of the design elements become expressed in the next version of the custom-built tool? The process and the documentation of the process becomes integral to understanding the evolution of these systems – and their preservation.

Bill Hearn, who built the voltage-controlled video synthesizer the Videolab in 1975–76, describes how he felt once he realized that a certain machine could have been made better: 'Right after the machines I made for you [Woody Vasulka] and Jack Brice, I was a little ashamed of the way they were put together' (Hearn 1977–78). Hearn envisioned a new device, and to do so, broke down the process of what it was to be used for:

> But a funny thing happened. Bill Etra and Steve Rutt called me on the telephone [...]. Bill had started out saying he wanted a voltage-controlled colorizer. [...] And something snapped in my head and I said, 'What you guys really want is a synthesizer that has all these things and is capable of doing all the standard special effects.' So I sat down and [...] it was really a terrific thing to think of, to reduce everything in video to a few little modules that could be interconnected and produce all the standard effects. I had to ask myself, 'What do you do when you make a wipe? What do you do when you make a key? What do you do when you make a split screen? What do you do when you do all these things?' (Hearn 1977–78)

Thus older machines are often incorporated into newer machines – another form of adaptation. Engineers' processes are not that different from those of the artist when it comes to conceiving tools. The goal is to build in possibilities and options to be creative. Bill Hearn, in discussion with video artist Woody Vasulka, discussed artists' use of the Videolab and ideas of aesthetics:

> Vasulka: Have you ever written about aesthetics? Even if it was a total failure or a total success? [...] Something that you described how the tools could be used?
>
> Hearn: I have recommendations in here (the user's manual) [...] esthetically I say really simple things like, 'If you're going to colorize, don't run the saturation up too high, mix in the original image and overlay the color on top of it'. That's the kind of advice I give in general. [...] I think about the person who made organs for Bach, the person who designed and built them. And what I want to do, what I really lust after is to make machines that are so clear to a creative person and give them so many possibilities that they can use them. That it gives them freedom as an artist. [...] It's like putting the keys on an organ or how many pedals, or how many stops, how close they are to the keyboard. What's the most elegant way of giving tool control to the person who's using it. (Hearn 1977–78)

Ideas for new machines started from other preexisting machines. Machines could be added to or converted to something else – expanded on, upgraded, evolved. Modification to machines was common, especially since these artists wanted to push the images to new parameters. Carl Geiger, one of the founders of the experimental video center Synapse at Syracuse University in the 1970s, speaks of 'retooling' machines and describes ways to hack their production equipment to make it produce 'the psychedelic, the alternative, the non-traditional video images':

> We had a portable studio – it was actually black-and-white cameras but the switcher was color – and it had a colorizer in it. We were always hacking that. We learned that you could push two buttons at once on the different busses. It wasn't supposed to happen and there was a little thing that locked it out, so you couldn't do that. But you could

> actually push on this one little mechanical thing that prevented you from pushing two buttons – so that you could push two buttons. And then we learned that if you took the preview output and put [the signal] back into the inputs and then mixed it with itself, you would get a delayed stutter effect that was cool. Me especially and others were very interested in, you know, in effect, the psychedelic, the alternative, the nontraditional video images. The images that were, you know, more vivid and intriguing. So we were off and trying to hack different things in different ways. (Geiger 2011)

And there was a basic knowledge that this hacking process Geiger describes above was not legitimate nor totally accepted:

> You know there were engineers that would maintain our studio sometimes, and we had to hide that kind of stuff. We were hacking the equipment, and we would have to hide it so they didn't know what we had done. And then if it broke we'd have to, like, get all that stuff off of there and so that they would fix it and then we could do it back. (Geiger 2011)

Many early practitioners modified black-and-white cameras and decks to allow more control over the image or to create certain types of effects. And some of the distributors of video equipment in New York City, like CT Lui and Technisphere, also offered modifications to equipment. Technisphere offered six modifications to the Sony Rover open-reel ½" portable video system. These modifications included the following: installing camera target level control; installing a mike/line switch; installing manual audio gain control; converting a battery meter to a battery/audio meter; installing coaxial video output; installing coaxial video input with internal-external switch.[3]

Dave Jones, engineer-designer, describes the changes and modifications he made to cameras and decks in the 1970s:

> Probably one of the most common [modifications] that I did was a sensitivity adjustment on the cameras [like portapack or Sony 3400 cameras]. The cameras were not very low-light cameras. [...] The lens was the only adjustment you had. So a lot of times, you tried to go into a low-light situation and they would amplify to a point of just being noise. Or you went into a bright light area, and the automatic gain controls just shut everything down. So in some of those cameras, I put in a switch and a knob that'll allow you to override the automatic gain control and just manually control the setting for the two. You have to remember, this was the days of tubes in the cameras for the pickup, rather than solid-state devices like CCDs. So you basically had this glass tube with a voltage applied to the back. And as the voltage changed, it changed the sensitivity of the pickup. So I was able to tap into where that voltage was, and put it on a knob to allow somebody to basically adjust the sensitivity of the camera. (Jones 2007)

And modifications were also done to cameras to switch the raster up and down. Jones describes a simple adjustment to modify the camera to 'flip' or 'reverse' the video image:

> The pickup tube inside the cameras worked just like a monitor worked. It had a set of coils around it. And one set of coils was the horizontal, one was vertical. So it was very simple to reverse the wires for that and flip the image up and down or flip it side to side. [...] often, it would be a situation where somebody might hang the camera from a ceiling, and instead of the image being upside down, they could then switch the flip and it would be right-side-up again. You know, where it was typically compensating for the way that the camera was mounted, to get back to a normal image, rather than just flipping it, keeping the camera upright and flipping the image.[...] Some people would do it as a way of flipping an image, and might have multiple cameras in a shoot, where one image was normal, one was reversed. The Videofreex did a fairly well-known video back then where they used two images; one was normal and one was reversed. They were doing a kind of a mirroring effect. David Cort did it a number of times in some videos. (Jones 2007)

The Experimental Television Center's combined tool set created a meta-tool for users to experiment with. ETC's studio of multiple custom-made tools was in tandem called the 'System', which was integrated in such a way as to allow artists to combine machines to create their own version of this system. There was a matrix that allowed one to feed any machine into another, and so the entire studio was developed as a modular system that was completely patchable and allowed you to treat it like a database. Peer Bode speaks of this structural conception of ETC's studio System:

> I give Ralph [Hocking] a lot of credit for his vision about experimental structures and making it so that you could ask questions and try things out easily. So that these things of modularity and the matrix – I don't know, they went through so many different matrix panels, so that you could get things in and out of devices without rewiring each moment. And that actually made a lot of things happen. (Bode 2012)

Of course, these image-processing systems, such as the one at ETC, were complicated to use, and it took a certain amount of time to master and produce anything original using them. Bob Snyder, founder of the Sound Department at The School of the Art Institute of Chicago, worked with the Sandin Image Processor (IP). In fact he hired people to build one for his own studio, since he felt he was not part of the movement to 'heat up the soldering iron and go for it, even though that was very much part of the ethos of the times' (Snyder 2003). But he did recognize the importance of having unlimited access to the Sandin IP to create work, a concept of open-ended artist access that Ralph Hocking of Experimental Television Center endorsed as well. While the ETC had a vibrant artist residency program, Hocking always encouraged people to create their own

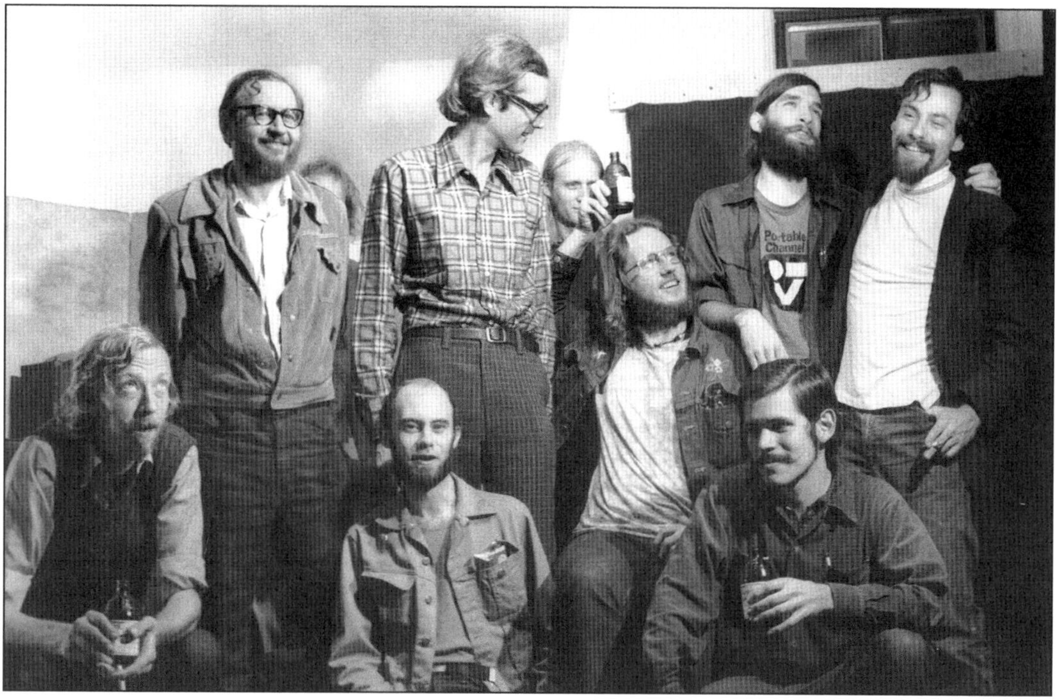

Figure 2. Tele-Techno Conference participants at Lanesville TV, 1975. Seated from left. Ken Jesser (Intermedia Art Center), Kevin Kenney (MERC), Dave Jones (ETC), Paul Lamarre ((Textronix). Standing. Don McArthur (ETC), Richard Monkhouse (EMS), Bill Claghorn (Adwar Video), Carl Geiger (Synapse), Chuck Heuer (Portable Channel), Chuck Kennedy (Media Bus). (photo & courtesy. Chuck Kennedy and the Videofreex Archive).

home studios as well, where they could use the machines daily and learn to play them like real instruments. In response to a question from interviewer Mona Jimenez about access, and likening the process of working with these machines to 'learning how to play an instrument', Snyder states that:

> I think that you've got to remember very, very few people ever had the kind of access you just described. Only us, you know, only us macho guys – who could commandeer the studio all summer or, you know, could build one and have it in our house – ever had the opportunity to even try to do that. Students got a little time and there were some brilliant students who did some remarkable things within the really limited amounts of time they had, but that's a really good point. I think you're absolutely right. You'd have to be very interested in using the instrument and have the time and the resources to really be able to tweak it and work on it for a long time. (Snyder 2003)

It must also be noted that while many women became videomakers, the custom tool design arena remained heavily male dominated. There were some exceptions, such as Sharon Grace (San Francisco), who built Paik/Abe Video Synthesizers under the tutelage of Shuya Abe, soldering boards and putting machines together; Barbara Sykes (Chicago), who built her own Sandin IP for her own use; and Louise Etra (Los Angeles), collaborator with Bill Etra and Steve Rutt, who had input into the functionality and actual design of the Rutt/Etra Video Synthesizer, among others. Mostly though, building machines remained a 'macho' activity with a limited number of women participating in it. Bob Snyder's comments resonate in terms of representing the gender politics of the period.

Snyder commented that it was hard to recreate anything with the Sandin IP, saying he hadn't the 'faintest idea' how he got certain effects: 'I just kept turning knobs until it looked right and then I stopped'. He worked with both music and video, not separating the two. Instead he developed a systematic way to approach working: 'The design of [the Sandin IP] was based on a kind of metaphorical transformation of modular analog sound synthesizers. [...] So rather than having the sound and the image be the same signal, they were different signals with the same control structure'. As a result of these kinds of experiences, Snyder considered the Sandin IP an instrument:

> I viewed it as an instrument and I viewed the monitor as an instrument. You know, I mean, one way to define an instrument is that it's a device. [...] An instrument is a – I mean not a scientific instrument, but a musical instrument – is a device that produces patterns of energy. And I very much viewed an image as a pattern of energy. I think that was the idea. Its color, its shape – all of those things to me were, you know, were like music. [...] There wasn't a strong distinction. (Snyder 2003)

Beyond questions of access, many artists in the 1970s were committed to the idea of needing to understand the creation of the tools they were using to be able to work with them most successfully. As well as understanding circuits, artists wanted to build their own machines for their own studios, and be a part of the 'heat up the soldering iron and go for it' ethos that Snyder spoke of. (See Color Plate 38.)

The wealth of tech in upstate New York

An example of this kind of commitment was the different ad hoc groups in upstate New York sharing technical information about building tools. Telephone exchanges in 1975 called the 'Tele-Techno Conferences' were inaugurated by Lanesville TV to encourage participants to share technological discoveries between various video working groups in New York State (supported by the New York State Council on the Arts – NYSCA).[4] The participants included Synapse, ETC, Portable Channel, Media Equipment Resource Center (MERC) and Lanesville TV.[5] Designer Dave Jones started something called the

'Tuesday Afternoon Club' in Barton, New York, near Owego, in 1982–83. Jones taught a group of artists to build circuit designs for their own machines, and to help develop prototypes created for the Experimental Television Center. The artists who participated included Peer Bode, Matt Schlanger, Barbara Buckner, Mimi Martin and Neil Zusman. Some of the artists went on to develop their own machines, and their understanding of systems and design influenced their artistic practices and determined their future directions. This idea of building your own circuits expanded from the 'Tuesday Afternoon Club' to the creation of a kind of open-source circuit board project through ETC called the 'Four Board Project', making boards available to artists to build their own systems. Dave Jones commented on the practice of sharing and building tools:

> Actually, within that same vein, there was a community of people who were building tools in the '70s. [...] That was a community of designers not so much copying each other's exact circuitry, but sharing design ideas with each other. It was concepts of 'if you take a keyer and do this to it, or run that into that, and wiggle this around.' [...] It was more general, conceptual conversations. 'Have you heard about this integrated circuit? Have you heard of this part?' – things like that. It was an entire side community in the video art world.
>
> [...] Also, in the early '80s, there were a number of people who were either a part of the TV Center, or were in the area and used the TV Center. I had moved to a house with a lot of empty space in it, so every Tuesday those people would come over [The Tuesday Afternoon Club]. Each person had a different thing they wanted to do; since they knew about the equipment at the Center, they wanted a keyer or a colorizer – they'd have a piece of equipment in mind, or an effect they wanted. Or I'd have a new idea and say, 'Well, I've been thinking about this or that,' and someone would say, 'Yeah! I want to try that'. So I would write up the schematics. A couple of them needed to be taught how to solder, so that was part of it, while others had done soldering so it was more advanced: 'Put this here and connect that there. Once that's working I'll hook it up to the oscilloscope'. So three or four people would sit around building circuits for themselves, slowly trying to build their own image-processing systems; meanwhile I was getting to try new things out. [...] It was a mixture of things: them learning and getting a circuit in the end, and me getting them to try out new ideas and building them for me while making another 30 or 40 dollars for the day. I think we met every Tuesday for a few months. (Jones 2005)

Matt Schlanger, artist and designer, successfully created his own studio much like the one at ETC. He describes the 'The Tuesday Afternoon Club' sessions:

> One of the original philosophies was that people would build their own. This System [ETC's system] is a prototype for individualized systems like this System, right?

> I'm Ralph [Hocking]'s poster boy here, ok? I did it. [...] A lot of the conversations on Tuesday Afternoons and not Tuesday afternoons with Peer in the room and others was about real time. Was about our process. Was about image processing as a movement which maybe never really had its voice the way it should have, but there were tributaries of philosophies around image processing, and part of it was process-oriented videomaking. Part of it of course which you know about was the signal processing and building your own machines [...] and embracing circuits and circuit design, and embracing the notion of the artist taking control of the means of production and doing things with it that are different than broadcast goals would dictate. (Schlanger 2012)

The ETC System was based on real-time processes, and the production of the machines in that System was based on 1960s audio synthesis processes. The development of these analog custom tools can be likened to building instruments, each machine having its own sensitivity and individual quirks. The analog devices were used in real time, which created a performative approach to working in the analog studio that was much like performing with musical instruments. The System was the instrument that needed tuning and the results were recorded on videotape.

Building tools also required structural thinking and designing, a kind of choreography of the machine, if you will. Peer Bode speaks of these benefits and of thinking about a system's infrastructure:

> Ralph [Hocking], here and there, got equipment that he bought really cheap, like racks – racks to put boards into that were a standard. And those racks probably were, like, $500 apiece new, and I think he got them for a dollar or two. And they were all in the back; they all had all the wire pins and they were all wrapped, connected, and you could disconnect – the first thing we had to do was disconnect them and just make them basic shells again, basic empty racks, so that we then would put stuff in. So all the stuff we were doing, pretty much [...] was also going into these kind of rack structures that Ralph had provided, that seemed like a reasonable kind of infrastructure idea for how to connect these things up. (Bode 2012)

There was a culture of scavenging used materials. This kind of ingenuity led to adaptive thinking and the creation of new ways to use these parts. This was a pre-eBay period of time, but tool designers were then very much tuned into ideas of repurposing surplus. IBM was in Binghamton, where the Experimental TV Center was, so there was lots of surplus.

Much like Dan Sandin's 'Distribution Religion', the 'Four Board Project' at ETC was produced as a means to disseminate information, to build and make available some basic image-processing circuits, and to enable artists to begin to build their own systems. The 'Four Board Project' was sponsored by a grant from NYSCA. The project also helped update the ways these circuit boards were produced.[6] Dave Jones was the lead designer on the project:

> The TV Center thought it would be great to have these types of image-processing devices printed as boards. Then we could make multiple copies – the Center could have one, I could have one, and different artists who worked at the Center could build their own copies. That was what became known as the Four Board Project: a multi-channel colorizer, a keyer, an oscillator and a sequencer. [...] And again, everything was voltage control. [...] I kept that going with the Four Board Project and everything was voltage controllable. We built a bank of oscillators for the TV Center, which are also still there. In fact all of the things we built with the Four Board Project are in the studio and used every day. (Jones 2005)

Another information-sharing session occurred around computer programming activities and was instigated by Ralph Hocking. Computers evolved into the Center's System over a number of years. The first ones had limited capabilities, and were slow, hard to program and generally difficult to use: 'Computers were not really capable of generating images in real time, or manipulating images beyond putting out some control voltages to turn knobs or throw switches' (Jones 2005). (See Color Plate 42)

In what video archive researcher and curator Carolyn Tennant describes as an 'exchange of ideas within the laboratory where experimentation happens' (Jones 2005), the artists and designers who were hanging around the Center started to develop a more versatile computer. Paul Davis and artist Walter Wright put together an NEA proposal and got a grant to buy and build up the computer. Davis and Wright, along with Dave Jones, all became part of another information exchange called 'Computer Sundays':

> Then probably in the early '80s, Ralph [Hocking] went to Paul and said 'OK. Things have changed with computers. Now there are personal computers'. There were a number of different things that Paul had been playing with at the school, setting up labs, so Ralph charged him with the task of finding a computer that we could do things with. So Paul put together a computer and we started playing with it.
>
> Because it took a while to write software for it, Ralph would invite us out on Sunday afternoon to write a program, make a modification, or add some new boards to the computer. It became a regular thing [Computer Sundays]. Paul and I would get together and spend the afternoon trying to add some program to do new features. They'd found a graphics card from a company in California that was capable of putting out video, initially a very low-res graphics card; we did some interesting little things with it, but they were just big blocks of color. Then they found this Frame Buffer card from this company that could actually capture full frames of video, make different scales of images, or manipulate an image and do something to it with software. A lot of the Sundays were spent trying to get that to work, and then writing programs to grab and manipulate images. [...] So we'd have an idea of things or interfaces to add that we'd throw out, and Ralph would respond. It was back and forth between me,

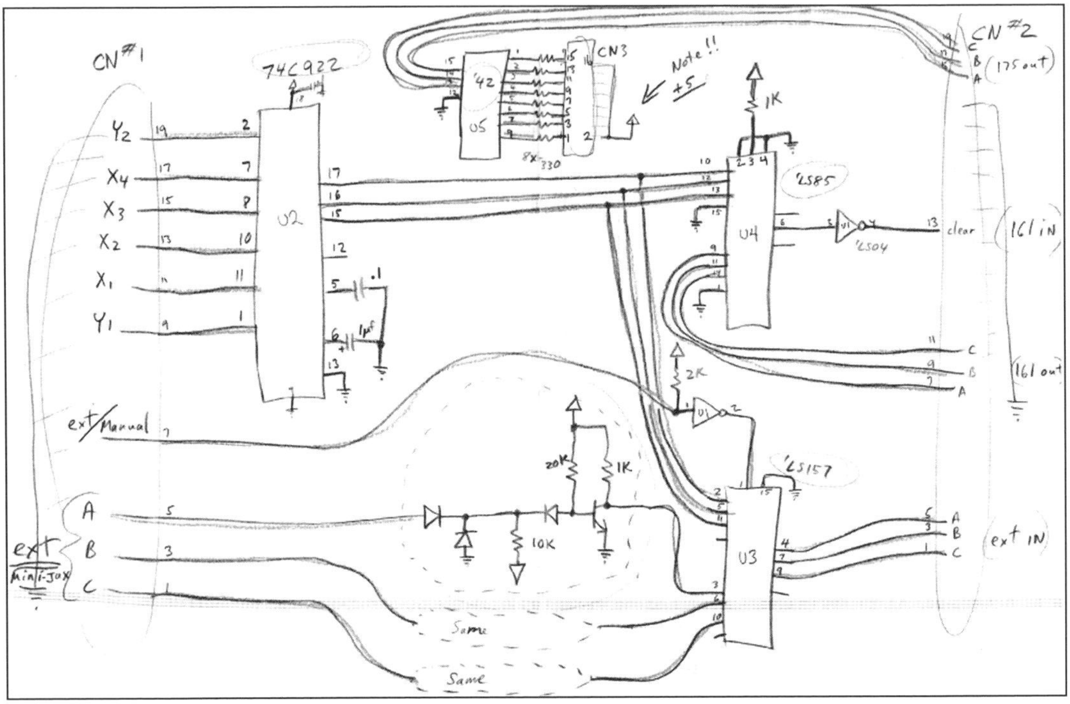

Figure 3. Schematic drawing for the control section of the Jones Sequencer designed by Dave Jones at the Experimental Television Center (1984).

> Paul and Ralph, playing around and trying to give the computer some features and character.
>
> [...] Whatever we worked on would either be based on Ralph's comments about something he wanted to do, or something we'd already done that he tried to use but wasn't happy with, and other times [...] we would have some other idea that none of us had tried before and so the next week maybe Ralph would be reacting to that. 'Oh can you do this?' 'Well, not quite. But we can do this instead.' Ideas came out of this back and forth, between what he wanted, what we saw possible, and then the results of when we tried to do it. And it didn't always come out the same way, anyways. (Jones 2005)

Because the time to program software for increasingly larger projects took longer than an afternoon, the Computer Sundays eventually stopped meeting. ETC had always been interested in developing computer interfaces with video image-processing systems across the upstate region, and worked closely with other media centers such as Media Study in Buffalo, Portable Channel in Rochester, Lanesville TV and Woodstock Community Video. As early as 1975 there was a concern on ETC's part to create a laboratory for tool development. Here is an excerpt from an NEA grant submitted in 1975 by ETC:

> There is at the present time a need for a center which serves as a focal point for technological developments in the fields of systems design, research and construction. The artist is probably best able to indicate his needs in terms of systems functions and a relationship between artist and scientist is essential for the development of systems. (ETC 1975)

ETC stands out as one of the experimental laboratories that focused on 'systems design and computer interface with video synthesizers' that was able to conduct ongoing research into this radical rethinking of media at the time.

Software – it is a good term

In this excerpt of a 1978 conversation between artist Woody Vasulka and designer Bill Hearn, questions were raised about storage, performance, and concerns with working between digital and analog:

> Vasulka: In digital there is a hope to get a score [similar to a musical score]. It would also be some sort of security. For analog, it requires high skill. It's not easy to work with analog systems. Some people get very good at it by practicing. You cannot just approach this tool and do a masterpiece. It in fact takes years of experience to use analog devices. So that digital has some hope of having a library. It suits this easy access to a library because it's possible. That is one advantage which digital [tools] will have [...]. These express very much the personality of the person who created them. [...]. There's only tradition's experience, and skill will die with people that die. The operators or artists that master it. It wouldn't get passed on as a score.
>
> Hearn: But a patch, all you have to do is take a picture of it or you can describe it digitally. It would be easy to encode a patch. I think that might take, maybe, only 200 bits of information to code each of these patches. Suppose more, suppose a thousand bits. You know, that's nothing.
>
> Vasulka: That's an interesting thought. You had better add a score module and then we can talk about something that can pass on this kind of experience. But it's very hard to expect that knowledge be accumulated. That's something for you to think about. How to pass it on. How to assemble a library that is craft oriented. A library that is of the craft of this tool.

Hearn: Software.

Vasulka: It doesn't have to be called software. But it's something like software.

Hearn: It's a good term. (Hearn 1977–78)

And as digital capabilities continued to emerge with the evolution of computers, software creation and digital tools were developed between collaborating artists and programmers. As artists crossed from analog into digital it was evident that the computers were not fast enough to process the huge amount of information that real-time video demanded. And artists were anxious for real-time video processing. As of the mid-1990s, programs were starting to be developed that could keep up with the processing time needed to produce real time. Artist Steina Vasulka worked closely with software designer Tom Demeyer to create Imag/ine in 1996–97. Image/ine was the first open-source software program for processing real-time video prior to others like Isadora and Jitter, which looked to this software for their development.

Steina stubbornly stayed 'an analog person' avoiding the frustrations and slowness of early digital work: 'Until I got to STEIM in Amsterdam. I was there as an artistic co-director and as I stepped into STEIM they said that they were making video software in honor of me. Which was very nice of them' (Vasulka 2011):

> In November of 1996, I arrived at STEIM[7] as the artistic co-director (with Michel Weisvisz) in Amsterdam. I had met Tom [Demeyer] previously, and of course knew about Big Eye, which I believe is the first motion tracking software written.[8] I was told that Tom was writing video software in honor of my arrival. I welcomed the endeavor, there could definitely be some use for primitive video software, though I knew that video had way more information than could be moved real time by software. [...] How was Tom going to write a real-time video in software, when we had in Buffalo in the late seventies built a huge hardware system for the same purpose? (Vasulka 2010)

Tom Demeyer, author of the Image/ine software, crafted the software for Steina. He felt their collaboration was productive:

> The development process became very creative and fruitful when Steina took up residency as artistic director of STEIM. She was of course very experienced with the analog machines, was able to guide and coax me in various ways, had a personal interest and an artistic input that really made it a 1+1=3 collaboration. (Demeyer 2010)

Steina and Tom's relationship matured over time:

SV: I was very condescending. Because I knew it couldn't be done. There was no way you could write real-time video, a real-time performing machine in the world because real time was not [...] there was no computer that was fast enough and there was no way it could be done. And he started showing me this and that and it was kind of OK. It was keying and some kind of colorizing. And then he showed me displacement.[9] [...] And I just wired up. We went through it and I think it was the first software – digital real-time video software. (Vasulka 2011)

TD: No, our ideas were not fundamentally different. Details of implementation, relative importance of features and functionality, sure. As a matter of fact, I remember us discussing how incredibly similar the steps in development of the different techniques and effects seemed to be. Hers a long time before in analog hardware, mine digitally, but seemingly following the same track. (Demeyer 2010)

SV: And it behaved like an analog machine but it was really MIDI controlled. And MIDI controls replace the famous knobs of analog devices – knobs and sliders. So I was in my element and it's still a wonderful machine, I use it a lot. So that's about digital. [...] You could use the property of the signal, of the electronic content to express your aesthetic desires. (Vasulka 2011)[10]

Starting with 160 x 120 @ 15 fps the resolution for Image/ine soon advanced to 320 x 240. Currently it is capable of running HD at 25 fps using limited effects. But as Demeyer said: 'Good people use what's available and aren't phased by technical limitations' (Demeyer 2010).[11]

By the late 1990s computers became faster and because of 'the third generation PowerPC computers were fast enough to do quarter resolution video in real time' (DuBois 2012). This speed issue had been frustrating designers since the 1980s. The software program Max enabled different programs to be developed: softVNS, NATO.0+55+3d, and Jitter. The flexibility of Max allowed various approaches. This was because of its 'data neutral paradigm' where anything was permitted and every object used in Max became equal. This concept of data neutrality can also be thought of as signal flow and voltage control, no matter what is generating the signal. And as new media artist and programmer R. Luke DuBois said: 'And so that voltage control thing was implicit in the semantics of the system from the beginning. Because everything can be something you experience or something that controls something you experience' (DuBois 2012).

So the video thing, we were waiting for [...] we started working on Jitter – when was it – probably 2000. And it took about two-and–a-half years to get working. And our alpha tester, our baseline tester was an artist in NY named Toni Dove. Toni Dove was working on a piece called *Spectropia* (2007) and we used it as a benchmark. And

> basically the theory was if we could do her wacky piece then it would be good enough for everyone else.
>
> [...] Toni Dove is a filmmaker – but there is a trick: she performs the editing of her films onstage. So she'll shoot like a full budget science fiction film – and shoot every shot six times, from two different persons' perspectives, close, medium and wide. And then onstage she has a motion-sensing instrument. She waves her hands and based on how she waves her hands, she can cut between the male perspective and the female perspective, go forward or go back, get inside people's heads or whatever. And she has done them as installations and she has done them as performances. [...] But we sort of wrote Jitter for her. The assumption being that if it could do what she needed then it would be fine for all the VJs that share on a Sunday night. [...] Because she needed to basically do four channel standard definition cutting in real time, controlled by a computer vision input, so she needed a camera and four movies all going at once plus sound, right? All scrubbed at 30 frames a second. So we developed it and it was her benchmark. (DuBois 2012)

The creation of the video software packages Image/ine and Jitter were built around the work of artists Steina and Toni Dove. While these women did not participate as much in the actual hands-on design of these tools, their performance methodologies and their ways of producing their video work shaped the conceptual development of each software. In fact, they served as the 'benchmarks' for the outer limits of the software.

These examples provide just a small discussion of the ways that tool design and collaborations can happen. Today, the boundaries of artist/engineer are perhaps blurring more and there are many more artist/technologists programming and building tools. And we will see how these changing skill sets will reshape tool production and collaborative processes for the future.

Make the generation happy

The impulses of artists and engineers to build machines – be they analog or digital – have remained the same: to inspire a broader public to engage in a creative act. From the early 1970s, video artists were teaching and helping students how to think critically, and creatively and to build their own systems. Among others, Ralph Hocking was at SUNY (the State University of New York) Binghamton, the Vasulkas were at SUNY Buffalo, Dan Sandin at the University of Illinois at Chicago Circle Campus and Nam June Paik with Shuya Abe at the California Institute of the Arts (Cal Arts).

Between 1971 and 1972 Paik and Abe took on seven students at Cal Arts, challenging them to build their very own Paik/Abe Video Synthesizers. Sharon Grace, a young artist at the time, partook in this extended workshop. Grace then also became a close collaborator with Paik and Abe, traveling around the country with them, soldering and helping build the machines they left seeded all over the country: at WNET, WGBH, the Experimental TV Center, Cal Arts and MIT. She was a quick study and Abe-san helped coach her with her soldering skills. The three became a production team. Eventually they felt their work was done and Abe-san returned to his family in Japan, while Paik and Grace went onto other projects. In an interview Sharon Grace commented on the sentiments behind Paik and Abe's early 1970s workshops, where they trained students to build the Paik/Abe Synthesizer:

> Whenever people asked them why they had done this, they would say that they were worried that the American youth were going to self-destruct on psychedelic drugs and this was their alternative to it. And the kind of sorrow and horror and grief that was internalized by everybody [from the Vietnam War]. [...] They would always say because the American youth was so depressed that they wanted to make them happy. [...] They would say that over and over and over again. [...] And they really did want to make the generation happy. (Grace 2012)

The youth of the United States were certainly mourning involvement in the Vietnam War in the early 1970s. Perhaps these early video tools had a more profound effect in changing today's communication dynamic than we can ever give them credit for. We were lucky to have had great teachers. Thank you.

References

Bode, Peer (2012), 'Interview by Kathy High and Mona Jimenez', June 7, Owego, NY: Experimental Television Center Archives.

Demeyer, Tom (2010), 'Tom Demeyer and Lenka Dolanova' [email exchange].

DuBois, R. Luke (2012), 'Interview with R. Luke DuBois by Mona Jimenez', April, 18, New York University & Owego, NY: Experimental Television Center Archives.

Geiger, Carl (2011), 'Interview with Carl Geiger by Mona Jimenez', February 5, Lafayette, NY & Owego, NY: Experimental Television Center Archives.

Grace, Sharon (2012), 'Interview with Sharon Grace by Kathy High', July 12, San Francisco, CA, Averill Park, NY & Owego, NY: Experimental Television Center Archives.

Hearn, Bill (1977–78), 'Interview with Bill Hearn by Woody Vasulka', Experimental Television Center Archives.

Jones, Dave (2005), 'Interview with Carolyn Tennant', *The Tuesday Afternoon Club and Computer Sunday at ETC – Ways Tool Design and Information Was Disseminated*, Owego, NY: Experimental Television Center Archives.

—— (2007), 'Interview with Kathy High and Mona Jimenez', July 31, Owego, NY: Experimental Television Center Archives.

Manoury, Philippe, Puckette, Miller, Lippe, Cort and Tennant, Carolyn (2005), 'Jupiter Interview: Philippe Manoury, Miller Puckette, Cort Lippe, Carolyn Tennant', June 11, Buffalo, NY & Owego, NY: Experimental Television Center Archives.

National Endowment for the Arts (NEA) (1975), 'Computer Interface with Video Synthesizer: Proposal – BIBLIO 2484', Owego, NY: Experimental Television Center Archives.

Paik, Nam June (2011a), 'Mr. Abe – The Greatest Doctor for Me', in *Paik/Abe Video Synthesizer: As freely as Picasso, as colorfully as Renoir*, Seoul, South Korea: Nam June Paik Art Center, pp. 14–21.
——— (2011b), 'Versatile Color TV Synthesizer', in *Paik/Abe Video Synthesizer: As freely as Picasso, as colorfully as Renoir*, Seoul, South Korea: Nam June Paik Art Center, pp. 8–10.
——— (2011c), 'Video Synthesizer Plus', in *Paik/Abe Video Synthesizer: As freely as Picasso, as colorfully as Renoir*, Seoul, South Korea: Nam June Paik Art Center, pp. 11–13.
Park, Sang Ae and Lee, Ki Jun (2011), 'Paik/Abe Video Synthesizer', in *Paik/Abe Video Synthesizer: As freely as Picasso, as colorfully as Renoir*, Seoul, South Korea: Nam June Paik Art Center, pp. 22–39.
Quasha, George and Stein, Charles (2009), *An Art of Limina: Gary Hill's Works and Writings*, Barcelona: Ediciones Polígrafa.
Rutt, Steve and Etra, Bill (2007), 'Interview with Steve Rutt and Bill Etra by Mona Jimenez', December 7th, New York, http://www26.us.archive.org/details/BillEtraSteveRuttInterview. Accessed June 12th, 2012.
Schlanger, Matt (2012), 'Interview with Matt Schlanger by Mona Jimenez', May 31st, Nyack, NY & Owego, NY: Experimental Television Center Archives.
Snyder, Bob (2003), 'Interview with Bob Snyder by Mona Jimenez', May 10th, Chicago, IL & Owego, NY: Experimental Television Center Archives.
Vasulka, Steina (2010) 'Steina Vasulka and Lenka Dolanova' [email exchange].
Vasulka, Steina (2011), 'Interview with SteinaVasulka by Mona Jimenez', January 19th, Santa Fe, NM: Experimental Television Center Archives.

Notes

1. Mechanical television, while popular in the US in the 1920s and 1930s, can be dated back to late 1880s.
2. Carolyn Tennant: 'While working as a researcher at Institut de Recherche et Coordination Acoustique/Musique (IRCAM) at the Centre Pompidou in the late '80s, computer programmer Miller Puckette developed a graphical programming environment that facilitated the work of composers using computers in their experiments with audio processing. Since its invention, the audio program known originally as Max has evolved into both a sonic and visual software program, Max/MSP Jitter, which today is available commercially for personal computing (Puckette went on to develop an open-source application in the '90s called Pure Data or PD). The program is widely used not only by artists working with real time audio and video signal processing but also as the engine behind countless interactive environments such as multimedia installations' (Jones 2005).
3. Undated product literature from Technisphere, ETC archive. Thanks to Sherry Miller Hocking.
4. Further discussion of the 'Tele-Techno Conferences' is detailed in Jeremy Culler's article, 'The Experimental Television Center: Advancing Alternative Production Resources, Artist Collectives and Electronic Video-Imaging Systems' in this book.
5. See http://www.experimentaltvcenter.org/tele-techno-notes. Accessed July 12, 2012.
6. Sherry Miller Hocking: 'In 1984–85 ETC, in conjunction with Matthew Schlanger, Connie Coleman and Alan Powell, documented the "Four Board Project". The intent was to make the documentation available at no cost for artists to use to build their own systems. We also discussed the idea of developing kits, but this proved too complex. Much of the documentation is now available online at the Video History Project.'
7. 'STEIM – the studio for electro-instrumental music – is the only independent live electronic music center in the world that is exclusively dedicated to the performing arts. Amsterdam, Netherlands.' From their website: http://steim.org/about/. Accessed July 12, 2012.
8. The Big Eye software was developed at the request of Fred Kolman, who had developed the tracking motion system with a hardware engineer, Andries Lohmeier, while a student at the AKI school in Enschede.
9. 'The concept [of displacement] is you build it up on the property of video. Which you have right from the get-go. Analog video, you have 0 intensity is black and 1 volt is white. And any gray scale in between is a fraction of 0 and 1. But not like zero and one like in digital, this is an analog curve. And this property makes it possible to key, because you key out first the darkest and then slowly up. Or you key from the brightest – you take out that particular part of the gray scale. And in displacement [...] you have one picture, you could say that's the top picture. And then under it you have another picture and it is the other picture's properties that get lifted up into the surface picture. So if the bottom picture [...] whatever is black there is no distortion and if it is white there is distortion. [...] But it's a displacement that made the brightest part of the picture lift up' (Vasulka 2011).

10. 'I was at STEIM for about a year and after I left Tom designed the graphic panels, which nailed all the loose ends into a system, and Sher Doruff wrote a very extensive instruction manual. STEIM offered the program for sale, there were many takers, and soon there was a user blog named LEV. Unfortunately, the site got infested with a brutal takeover by "Netochka Nezvanova" a software writer who build video modules (NATO.0+55+3d) as an add-on to the Max software. She would not tolerate any im/i discourse, and eventually killed the site (I think). Mark Coniglio was an artist-in-residence during my stay at STEIM. He became inspired to write his own video software, titled Isadora. Not long thereafter rumors of another Max video software turned to be Max/Jitter. With every new generation of MacIntosh, the software improved and expanded. Tom used the opportunity of OSX operating change to totally revamp the graphic part. At that point, it was renamed imX and became a freeware, which often is a death sentence for software. There is no denying that imX was crashing a lot and the new graphic interface was still undergoing changes. Apple's move to Intel has done very well for imX, and whereas Isadora, Jitter and Pure Data are well suited for musicians, imX seems to speak better to visual artists' (Vasulka 2010).

11. '[T]here had been two other attempts at making a video packet for Max. One was called VNS [softVNS] made by David Rockeby, an artist up in Toronto – the Very Nervous System. And the other one was called NATO [NATO.0+55+3d] and was developed by a very interesting consortium of Dutch men pretending to be a woman from New Zealand as a sort of performance art act. So it was a sort of hacker collective where they had a very serious activist bent. So the software package cost more if you were from a NATO country in protest of the bombing of Kosovo. You know all this kind of stuff. And the public persona was a woman named Netochka Nezvanova– which was the Dostoyevsky character. And they hired this woman whose name was Rebekah Wilson who was a performance artist from New Zealand to pretend to be this programmer. But actually it's a bunch of guys from audio – it was a bunch of men from the Netherlands. But they had this front woman who was doing this kind of stuff. And that turned out to be – well the company, and it wasn't even really a functional company, but they went under at about that time' (DuBois 2012).

Computer-Based Video Synthesizer System, ETC

Donald McArthur, Walter Wright and Richard Brewster

EXPERIMENTAL TELEVISION CENTER LTD.
164 COURT ST.
BINGHAMTON NEW YORK 13901
607-723-9509

Computer-Based Video Synthesizer System

This report summarizes the results of the research and development of a computer controlled video image processing and synthesizing device, a project supported in part by the National Endowment for the Arts and the New York State Council on the Arts.

Dr. Donald McArthur, Systems Design
Mr. Walter Wright, Software Development
Mr. Richard Brewster, Systems Construction

October 1977

The Experimental Television Center Ltd. is a not-for-profit educational corporation supported in part by the New York State Council on the Arts and the National Endowment for the Arts.

EXPERIMENTAL TELEVISION CENTER LTD.
164 COURT ST.
BINGHAMTON NEW YORK 13901
607-723-9509

As video image generating, processing and synthesizing systems become increasingly sophisticated, the problem of achieving maximum control over these systems must be addressed. This is particularly relevant to situations in which videomakers work independently and frequently as individuals in the creation of their works. Many video systems provide such a large number of image making variables that manual adjustments within the parameters of each control necessary to obtain desired structures and sequences is not always possible. The artist must then compromise the image to the system. Video synthesizer systems offer an enormous potential for intricate image constructions, but without appropriate control systems the individual artist may not be able to take full advantage of the system to achieve with accuracy the structures desired. Microprocessor systems, it was believed, could provide the necessary control for precision of image structuring if these computer systems could be completely dedicated to the processes of visual art making and be made usable by artists in direct ways.

The primary aim of the Computer-Based Video Synthesizer project was the research and development of such systems with capacity for direct use by artists in the production of independent works. A major design consideration in both hardware and software development concerned the establishment of a holistic system directly related to the requirements of individuals working in the electronic arts and usable by them in their personal work. It was considered important to reduce, as much as possible, the reliance by artists on outside technical support in the production processes because of the difficulties of communication and interpretation and the intimacy of the creative situation. The initial planning for this project began in 1975 with conferences at the Experimental Television Center involving Mr. Ralph Hocking, Mr. Walter Wright, Dr. Donald McArthur and Mr. Richard Brewster of Binghamton, New York and Steina and Woody Vasulka of Buffalo, New York. With support from the National Endowment for the Arts and the New York State Council on the Arts, the project was completed in the Fall of 1977; the resulting system is now operational at the Experimental Television Center and is available for use by artists through the production program at the Center.

An important philosophical consideration throughout the project concerned a humanistic approach to the design and utilization of computer and video systems technologies; one of the initial stages of this project involved the development of methodologies which would guide the construction of complex tools and systems dedicated to the needs of visual artists. To achieve this end, it was essential that artists, programmers and engineers work together in all aspects of the project, each group communicating from its own unique perspective. Artists helped to articulate and define the types of controls which they felt were important in image making. Engineers and programmers frequently introduced image making devices and control methods which had not previously been available; the structural and compositional potentials of these components were completely unexplored. In the design and construction of the hardware there were a number of specific objectives. A flexible and versatile system was important in order to provide artists with as many options as possible for image generation, processing and control. The present system is modular in design and permits the inclusion or exclusion of discrete components in the assembly of a system specifically tailored to meet the individual requirements of a particular artist or project. Modular and standardized design also allows for the future research and development of new components and the modification of existing hardware and software all of which can be incorporated into the present system with a minimum of effort. The system is capable of interfacing with many video and computer components an increasing number of which are owned by or accessible to individual artists and small arts organizations. For example, the system at the Center is compatible with the system of Steina and Woody Vasulka, and exchanges of software and hardware are possible between Buffalo and Binghamton. Video production requires a fairly powerful microprocessor system which is capable of efficiently handling the large amount of information necessary in the generation and control of image structures. The needs for a powerful system, also low in cost, had to be weighed against the factor of complexity since the system was to be used by individual artists the majority of whom had little or no prior experience with computer hardware or software. The 16 bit system as it was designed and constructed met the criteria of low cost, high power and ease of operation. The hardware made use of commercially available components as much as possible in the interests of efficiency of operation and construction and ease of duplication of the system by artists and arts groups. Many specific components and interfaces, however, had to be designed and constructed specifically for this project since they were either not available commercially or were too costly; many of the commercial components which were available did not meet the specific requirements determined by the nature of the application of the system.

The software development for this project also emphasized a humanistic approach to the use of microprocessor and video systems by artists. The goal of the software research was the development of an interactive language usable by artists. This language had to be understandable to artists so that they could address the computer directly, using language and concepts derived from the visual arts, without the necessity of translation into high level computer languages. The language had to be responsive to the needs of artists, enabling them to manipulate discrete elements of design and compositional structures. Further, it had to allow the artist to intervene at any point in the construction of the composition so details of compositional configurations as well as whole sequences could be easily altered. Precision was felt to be critical; the artist had to be able to develop and score the composition, store, run and edit it in a manner which insured its accuracy and repeatability. It was felt that the language should also provide for the option of programmed randomness and operate in either structured or random modes or a predetermined combination of both modes.

The computer-based video synthesizer system which is now operational at the Center consists of two sub-systems, the microprocessor and the video system and their interface. The computer section consists of a 16 bit DEC LSI-11 microprocessor, teletype and printer, dual floppy disk and 20K of memory. Components designed and constructed specifically for this project include the parallel interface, buffer memory, module to element bus, element bus, digital to analog converters, analog to digital converters and real time input. The video system includes a four channel analog colorizer with keyers, a 50 point switching matrix, spatial and intensity digitizer and a voltage control bank. The video system is modular in design; each of these components was researched, designed and constructed over a period of four years under the research and development program at the Center. Each of the video components may be combined with any other to form a system tailored to individual requirements; the video system may be operated manually or placed under computer control. This design consideration allows a maximum flexibility with a limited amount of equipment, permitting the same components to serve a variety of artists with different systems needs and experiences. Hardware design also permits manual interruption of computer processes at any point through the use of analog to digital converters and real time input. This feature allows the artist more complete control over all elements of the image and its temporal structures. Changes in composition may occur by software reprogramming or direct manual interactions by the artist or a combination of these techniques.

A more detailed description of the hardware aspect of this project is presented in the papers by Dr. Donald McArthur. Section I A provides an orientation to the system architecture. Section I B is a paper written from a transcript of a presentation by McArthur in Buffalo, New York in March 1977 for the 'Design/Electronic Arts' conference supported by the National Endowment for the Arts and the New York State Council on the Arts and sponsored by Media Study/Buffalo and the Center for Media Study, State University of New York at Buffalo. This presentation by McArthur was based directly on the research McArthur had done for the Computer-Based Video Synthesizer project.

The aim of the software aspect of the project was the development of an interactive language which uses concepts and vocabulary derived from the visual arts. It was anticipated that this approach would make the computer based video system accessible to a much larger number of artists than would a system which depends on the presence of a programmer to interpret the ideas and images of an artist into a computer language. Before any except the most rudimentary of programming could be developed, analysis of the fundamental elements in the composition of single images as well as their temporal structures had to be conducted. Identification and definition of these elements and the parameters of change within each variable were the initial steps. Within single images, discounting the time function, elements which were chosen included color field variables such as hue, saturation, chroma and intensity, form and shape variables including type of shape, position and frequency, texture and density. Each element has parameters of change which involve the temporal aspects of video. The methods of change involve problems of duration and sequencing with references to rhythmic structures.

As is noted in the papers by Wright, the software research is still in its initial stages and further explorations are necessary before the interactive language is fully functional. Several programs have been developed, one of which is analyzed in Wright's paper, section II A, which represents an incomplete stage of the language. Section II B is a transcript of a presentation by Wright at the 'Design/ Electronic Arts' conference in Buffalo in 1977; these materials are based directly on the research Wright had done for this project.

The computer based video synthesizer system is now operational at the Experimental Television Center in Binghamton, New York and is available to artists under the production program. As a greater number of artists utilize this system, each artist will be encouraged to articulate ways in which the system can be made more responsive. The results of this project have already indicated several important avenues for continued research, among them further and continued software development and the publication and dissemination of the results of the research to date. The computer based video system can serve as a model system; publication of research results will allow the duplication and modification of the system by other individual artists and arts organizations. Although the research to date has been specific to video, microprocessor systems are useful tools in many of the visual and performing arts, and a publication of this nature would assist many individuals from a variety of fields. A complete set of documentation has already been prepared; the next phase of this project, for which the Center is seeking support, includes the publication of these materials, including detailed schematic documentation. This publication will also include more theoretical papers, approaching the system and its applications from the points of view of aesthetics, physics, electronics and video and microprocessor technology. The aim of this publication is to provide specific and detailed information to permit duplication of the system and also introduce conceptual frameworks from which to view the electronic arts.

Design/Electronic Arts: The Buffalo Conference March 10–13, 1977

John Minkowsky

In March 1977, not many months into my tenure as 'Video/Electronic Arts' Curator at Media Study/Buffalo (MS/B), a regional center in Western New York, I was given the opportunity to organize a four-day conference, 'Design/Electronic Arts', to explore the collaboration of video and audio artists and theorists/engineers/technicians/scientists in devising new tools for creative imaging and sound. Two dozen prominent innovators in this field were invited to exchange ideas regarding the principles of design and practical information about it. While suggestions concerning the roster of participants came from many sources, the two most generous advisors were Woody Vasulka and Gerald O'Grady. Nearly all the sessions were conducted in a large auditorium at the Marine Midland Bank in downtown Buffalo. Funding came from the New York State Council on the Arts, the National Endowment for the Arts and the Center for Media Study (CMS) at the State University of New York at Buffalo (as a cosponsor).

I was founding Curator at MS/B from the fall of 1976 through the end of 1984, and organized 300 presentations and performances by videomakers and musicians, as well as touring exhibitions and residencies by artists to create new works and to design audio tools. I came to the position with a substantial awareness of the current trends in both video and electronic music, having arrived as a freshman at the State University of New York in 1971, just as CMS and MS/B were coming into existence. Every week, a different artist working in video was invited to make a screening/presentation at the University, and I attended them all, taking in the full range of video experimentation. I was also fortunate to have Stan Vanderbeek, Ed Emshwiller, Peter Campus and Nam June Paik as some of my earliest instructors in videomaking. Paik, in a short seminar, introduced everyone to the raster distortion of broadcast signals by placing powerful electromagnets atop TV monitors.

Buffalo was, therefore, teeming with electronic media activity by virtue of both the Center for Media Study at the university and the independent Media Study/Buffalo – both founded by Gerald O'Grady – with a particular interest in the development of new tools for electronic art, of which there were perhaps only two other locations in the country (the Experimental Television Center in Binghamton, New York, and the Electronic Visualization Center in Chicago [both well represented at the conference]). Buffalo was also the home of audio-tool designers Robert Moog and Harald Bode, and the computer music pioneer Lejaren Hiller, with whom I had various degrees of interaction. The arrival of Steina and Woody Vasulka to teach at CMS in 1974 proved to be the final key ingredient, for they had been the preeminent innovators in exploring

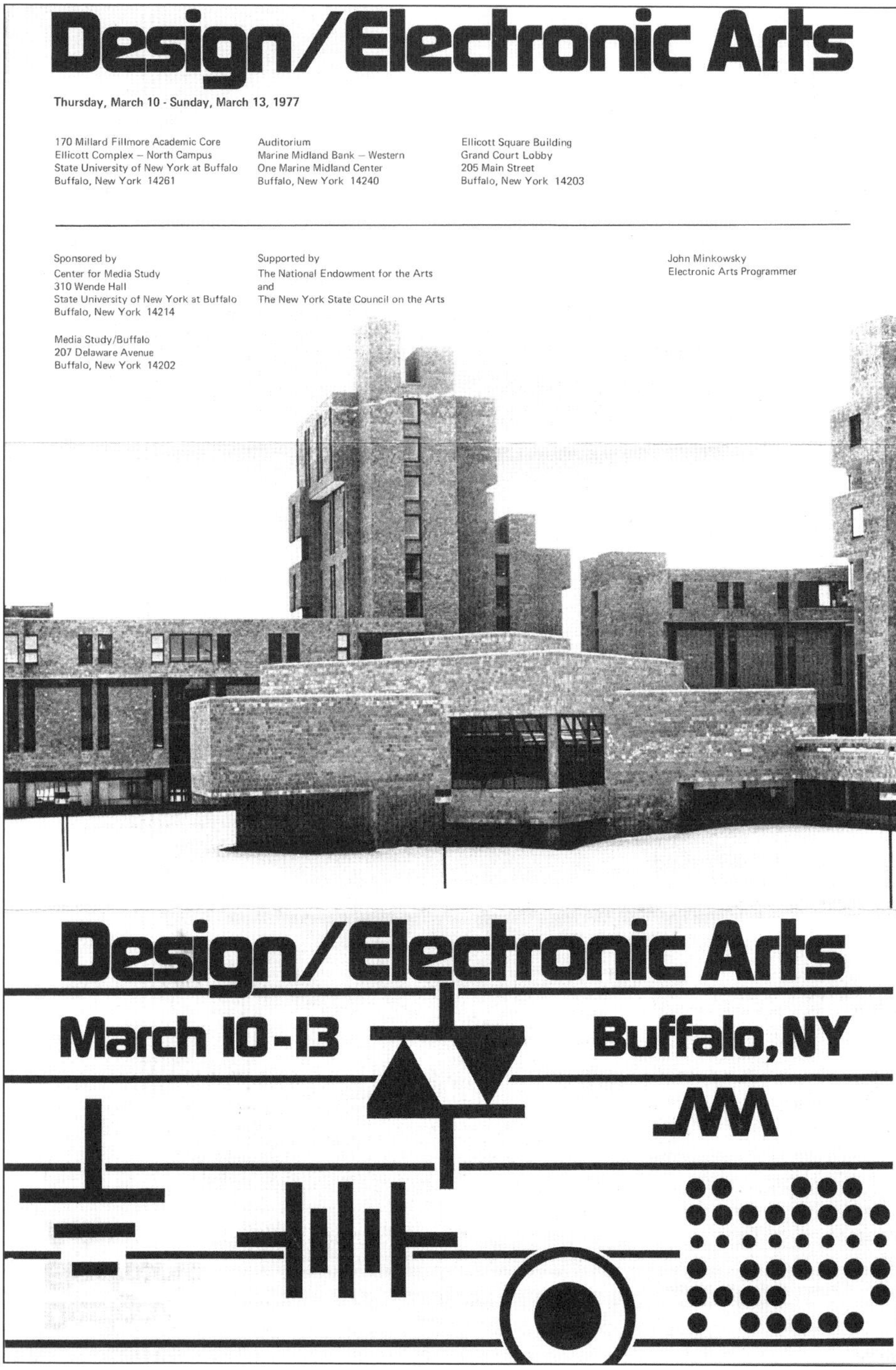

Figures 1 and 2. Poster for 'Design/Electronic Arts: The Buffalo Conference', March 10–13, 1977.

Design/ Electronic Arts

Thursday — March 10, 1977

6:00 PM
Fargo Dining Hall
Ellicott Complex – North Campus
State University of New York at Buffalo
Buffalo, New York 14261

RECEPTION
By Invitation Only

8:00 PM
Room 170
Millard Fillmore Academic Core
Ellicott Complex – North Campus
State University of New York at Buffalo
Buffalo, New York 14261

Self-Observing Systems: The Video Revolution in America
Gene Youngblood
Film & Television Study Center
6233 Hollywood Blvd. at Vine
Hollywood, California 90028

Open to the Public

Friday — March 11, 1977 — By Invitation Only

Auditorium
Marine Midland Bank – Western
One Marine Midland Center
Buffalo, New York 14240

9:30 AM
Introduction
Gerald O'Grady
Director, Center for Media Study
310 Wende Hall
State University of New York at Buffalo
Buffalo, New York 14214
and
Media Study/Buffalo
207 Delaware Avenue
Buffalo, New York 14202

10:00 AM
Technological Art: Some Problems
Ken Knowlton
Bell Telephone Laboratories
Room 2C 525
Murray Hill, New Jersey 07974

11:00 AM
The Geometry of Consciousness
George Chaikin
Brain Research Laboratory
New York Medical College
Flower Fifth Avenue Hospital
at 106th Street
New York, New York 10029

12 Noon
Visible Languages: Towards Diagrammatic Environmental Communications
Aaron Marcus
School of Architecture and Urban Planning
Princeton University
Princeton, New Jersey 08640

1:00-3:00 PM
Lunch

3:00 PM
Designing in Time: Common Principles in Visuals and Sound
Laurie Spiegel
175 Duane Street
New York, New York 10013

4:00 PM
A Description of the Landscape Within Which Computer Music Systems Are Designed
Joel Chadabe
Department of Music
State University of New York at Albany
Albany, New York 12222

5:00 PM
Applications of Speech Synthesis to Music
Joseph Olive
Bell Telephone Laboratories
Murray Hill, New Jersey 07974

6:00-8:00 PM
Dinner

8:00 PM
Generative Systems–The Missing Media
Sonia Sheridan
School of the Art Institute of Chicago
Michigan Avenue and Adams Street
Chicago, Illinois 60603

9:00 PM
Music with Roots in the Aether: Video Portraits of Composers and Their Music
Selected Short Subjects: (1) Cast and Crew, (2) Manners, (3) Are There Any Questions, (4) There's Always a Scandal, (5) Coming Attractions, (6) Intermission
The Feature: What She Thinks
Robert Ashley
Director
Center for Contemporary Music
Mills College
Oakland, California 94613

Saturday — March 12, 1977 — By Invitation Only

Auditorium
Marine Midland Bank – Western
One Marine Midland Center
Buffalo, New York 14240

10:00 AM
3-D Digitizing and Motion Description
James H. Clark
Information Sciences
University of California at Santa Cruz
Santa Cruz, California 95064

11:00 AM
A Description of the Anima-II 3-D Color Computer Animation System
Charles Csuri
Ohio State University
Research Center
1314 Kinnear Road
Columbus, Ohio 43212

12 Noon
Synthavision: An Overview
Larry Elin
Synthavision
3 Westchester Plaza
Elmsford, New York 10523

1:00-3:00 PM
Lunch

3:00 PM
How to Apply What You Know About Analog Electronic Art to Your First Computer
Tom DeFanti
Department of Chemistry
University of Illinois at Chicago Circle
Box 4348
Chicago, Illinois 60680

4:00 PM
The Image Processor: Design, Operation, Use
Dan Sandin
Department of Art
University of Illinois at Chicago Circle
Box 4348
Chicago, Illinois 60680

5:00 PM
Crosseye, General Motors – 1976, Re-Scanning of Eve-1 and Eve-88 and Glimpses of Other Entities in Process
Phil Morton
1839 S. Halstead Street
Chicago, Illinois 60608

6:00-8:00 PM
Dinner

8:00 PM
From Slit-Scan to Raster-Scan
John Whitney
Motion Graphics
600 Erskine Drive
Pacific Palisades, California 90272

9:00 PM
Art and Technology
Stan Vanderbeek
Department of Art
University of Maryland
Baltimore County
Baltimore, Maryland 21201

Sunday — March 13, 1977 — By Invitation Only

Auditorium
Marine Midland Bank – Western
One Marine Midland Center
Buffalo, New York 14240

10:00 AM
Pantomation
Tom Dewitt
111 North Pine Avenue
Albany, New York 12203
Phillip Edelstein
Technical Supervisor
Electronic Music Studies
State University of New York at Albany
Albany, New York 12222

11:00 AM
Digital Meets Video: The Best of Both Possible Worlds
Lou Katz
College of Physicians and Surgeons
Columbia University
630 West 168th Street
New York, New York 10032

12 Noon
Computer Control of Real Time Television Devices and Devices for Micro-Processing
William Etra
42 E. 23rd Street
New York, New York 10010

1:00-3:00 PM
Lunch

3:00 PM
A Computer-Based Video Synthesizer: Part I Hardware
Don McArthur
Experimental Television Center
164 Court Street
Binghamton, New York 13901

4:00 PM
A Computer-Based Video Synthesizer: Part II Software
Walter Wright
Experimental Television Center
164 Court Street
Binghamton, New York 13901

5:00 PM
The Territory, The Subject and The Transmission of Knowledge
Woody Vasulka
Center for Media Study
State University of New York at Buffalo
310 Wende Hall
Buffalo, New York 14214

6:00-8:00 PM
Dinner

Sunday — Open to the Public

Ellicott Square Building
Grand Court Lobby
295 Main Street
Buffalo, New York 14203

8:00 PM
PERFORMANCES BY THE SONIC ARTS UNION

Robert Ashley
The Great Northern Automobile Presence

David Behrman
Music for Cornet, Micro-Computer and Homemade Electronics
David Behrman - Electronics
Jim Horton - Computer Programmer
Gordon Mumma - Cornet
Kim One - Harmonic Changes

Alvin Lucier
Bird and Person Dyning (1975)
For Performer with Microphones, Amplifiers, Loudspeakers and Sound-Producing Object

Gordon Mumma
Passenger Pidgeon (1676-1976)

Co-Sponsored by Center of the Creative and Performing Arts, SUNY/Buffalo, The University-Wide Committee on the Arts, and Meet the Composer, a Project of the American Music Center, Funded by The New York State Council on the Arts the materiality of the electronic media using new instruments designed for personal analytical and expressive purposes.

The particular artist/designers who were invited to the conference were selected from a much longer registry because they brought with them a range of approaches, which were nonetheless interrelated, to each of the categories of work described below. But there were reasons why some were not present as well. For example, Ivan Sutherland told me that he was no longer interested in making such presentations, having done so many in the past. For still others, like Shuya Abe in Tokyo, it would have been prohibitively costly to fly to Buffalo given budget constraints. And still others, however enthusiastic, simply were faced with scheduling conflicts that prevented their attendance. All of the initial negotiations were conducted by telephone, and so no written documentation of these communications is extant, although handwritten lists indicate the gradual winnowing away of prospective speakers and presenters. Still, overall, I was satisfied that we had covered the bases with a convocation of brilliant practitioners that would generate a lively exchange.

My intention at the conference was to bring together representatives of three types: conceptual designers, or philosopher/theorists of art, technology and communications; tool and system designers and builders, or those actively engaged in constructing hardware and software systems to be employed by artists; and designers of images and sounds themselves who made use of these new tools. As might be expected, these were soft distinctions that began to dissolve from the outset. Throughout the conference, there were attempts to identify where tool designers and artists could be distinguished in the collaborative cross-fertilization between these traditionally distinct realms.

The nature of the collaborations between artists and designers could take a number of forms, among them the hardware or software designer creating machines and programs to specifically meet the needs of a particular artist's vision; the artist learning the potentials and limitations of a system from the designer in order to create works based on these criteria, and an open exchange of ideas that could result in image and sound forms unforeseen by those involved, and in the creation of new instruments toward these ends. An artist's desire to affect certain manifestations of her own ideas might prompt the scientist to envision fresh approaches to his own art of post-industrial invention and application.

Here, in Sections 1 – 4, I present the situation vis-à-vis electronic art circa the late 1970s and how it had come to be (1), followed by a list of principles of electronic tool design as enumerated by Dan Sandin for his analog Image Processor (2). There follow summaries of the talks, demonstrations and performances by each of the 24 presenters, and a final address by Woody Vasulka (3). Finally, I offer my brief perspective on the techno-cultural developments in the more than thirty years since the conference (4).

Most quotations are directly from the audiotape records of these proceedings. Those from other sources have been footnoted.

1. The state of the art: 1977

It had only been a decade since Sony mass-marketed the portapak, the affordable hardware that initiated the independent video movement in America. That was phase one. The second phase, within a few years, was the move by a handful of artists to invent specialized analog tools – video synthesizers and processors, colorizers, and multi-keyers, among others – that permitted a wide variety of modulation of video signal waveforms in real time to allow vast arrays of 'special effects' beyond those available in the standard television studio.[1] By 1977, video was in its third stage – the transition from analog to digital generation and control of the television signal to further extend the creative use of the medium. It was then that the first home computers appeared for purchase by the general public, based on the microprocessor technology that came as a result of integrated circuits acting as central processing units. The first attempts at interfacing digital and analog – converting digital information back to the analog mode of the television system and vice versa – were, at this stage, often dependent upon the mainframe computer systems available only at large industrial, research and educational facilities that could afford them, although this was rapidly becoming less the case.

This had been the trajectory of video art in its first decade – from a primitive portable straight-on recording instrument, to analog signal-processing tools that were often unpredictable in their behavior, to the beginnings of precision control with digital computers.

While this is a huge simplification of developments during the first decade of video tool exploration, it does point to the fact that 'Design/Electronic Arts' took place at a pivotal moment – the cusp of video taking its next step into uncharted territory. And it is essential to reiterate that the transition from analog to digital was very much a new direction in tool making that would have enormous impact on all that came after in the electronic arts.

2. Some principles of tool design

Throughout the conference, as designers outlined some of the principles upon which they already worked in the hardware and software they were developing, the two most operative words were 'real time' and 'interactive'. No one articulated his own criteria with such specificity as Dan Sandin in describing his design of the analog Image Processor (IP). I present Sandin's list as an expedient template in considering the many ideas that were later recapitulated by others in different ways. Although Sandin focused on his analog IP at the time of the conference, most of the design ideals he expressed would extend to its digital counterpart already under way.

Real time: 'Analog systems are so dumb they can't store any information,' said Sandin, 'so the stuff's got to come out as fast as it goes in. It is an inherently real-time system in most cases, especially with video' (Sandin and Morton 1977).

This was not the case with digital tools, where storage capacity of signal data was to a large degree fundamental to the ways they were to be employed. The particularities of the hardware itself and complexities of the software programming could create delays between one end and the other, or determine the durational parameters of the act of creation based upon the time required to compute a single image frame. There was, at the beginning at least, a trade-off between analog spontaneity and digital control. But it was not an either/or situation: some systems permitted both modes of real-time decision making and more contemplative modes of composition, the scores of which could be retrieved immediately upon completion of a work for evaluation and revision.

Interactive: Interactivity is part and parcel with real-time systems, and it became more a question of degree and type; for example, how easily the user could make the cognitive connection between her actions and the images or sounds produced, and to what extent repeatability of a particular visual/aural event was possible.

Rich feedback: The immediate gratification of witnessing the results of one's actions is closely linked with real-time interactivity. Optimally, thought Sandin, feedback should be as rich in character and detail as possible.

General purpose: Systems should be general purpose, ones in which the logical organization or architecture would not present 'walls' (limits of flexibility) to the user. These were opposed to special-purpose turnkey tools – one-trick black boxes which performed with the flip of a switch – that limited what Sandin termed 'modes of thought' encouraged by more generalized flexibility.

Modularity: A key factor here was constructing the machine out of discrete units capable of particular input-to-output processing that a user could easily interconnect to generate images of enormously diverse natures. Modular systems could reproduce functions of the special effects console as well as many more radical visual manipulations. And the interchangeability of voltage controls and image allowed one image to affect another. Furthermore, the neophyte could first master the small domain of this array of modules to simulate a special purpose piece of hardware, and then begin to combine it with others to discover new complexities as his sense of competence increased.

Easy to learn: Admittedly, Sandin noted, general-purpose systems often contravene ease of learning. Digital computers, which are always general purpose, require that the user could proceed only with a substantial body of knowledge, such as complex assembly languages, rules of programming, and the generation of subroutines. But here, Sandin thought, the tension between too many and too few options could be mitigated by finding 'a slice through these two worlds that gives you a tremendous amount of generality and still maintains a conceptual clarity or kind of hierarchy that allows one to start knowing very little, with the machine teaching the rest' (Sandin and Morton 1977).

High tactility: This was one means of easing new users into the system, a manner of developing a direct relationship between what the hand does and what the eye sees. The IP used patch programming – something like a superannuated telephone switchboard – to plug together modules in as simple or labyrinthine manner as desired, but other peripherals like joysticks were also employed. Working at 'touchie-feelie' inputs of this sort obviously constituted a different process than typing elaborate instructions at a computer keyboard to obtain results.

Portability: Portability is a relative term. Although cumbersome, the IP could be packed and transported to other sites for performances, and this was a common *modus operandi* of Sandin's compatriot Phil Morton to create his two-way interactive installations. Mobility of other large systems was clearly impossible, and even Don McArthur, who aspired to portability with his Computer-Based Video Synthesizer, had up to that point only created hardware roughly comparable in size to two refrigerators.

Low cost: Putting a tool in the hands of as many individuals as possible to explore at their own pace was one of Sandin's primary objectives, and he provided the schematics gratis to anyone who wished to construct her own IP at an estimated cost of $3000. Compare this figure with the $10 million machine available at Columbia University and the discrepancy is breathtaking. But, as many noted, the home computer would bring digital tools out of the institution and directly into the hands of artists.

Issues of safety: An assurance that, in their interaction, neither was the user hurt nor the tool damaged was a design consideration for Sandin, and it is not an irrelevant one when speaking of intimate and extended hands-on employment of a complex electronic instrument. Stephen Beck, the designer of the Direct Video Synthesizer at the National Center for Experiments in Television in San Francisco, never let anyone else use his machine for fear that they might cause damage that would require countless hours to diagnose and repair.

These, then, were some of Dan Sandin's criteria for an ideal electronic imaging tool.[2] In what follows, the manner in which other designer/artists shared them and envisioned additional ones will become clear.

3. Presentations at the 'Design/Electronic Arts' conference

Placing Sandin's criteria in a proper context requires a description of the conference proceedings and participants. They are here categorized under 'Theory', 'Music', '3D motion graphics', 'Graphic arts' and 'Design by collaborative teams for computer-based video systems', as well as 'Evening events: presentations, a mini-retrospective and a concert', and Woody Vasulka's summary comments.

Theory

By theory I mean that which offers conceptual models that are projective, prescriptive, or analytic, or otherwise describe unusual and original systemic paradigms, and three presentations – those by Gene Youngblood, Ken Knowlton and George Chaikin – fell within this category.

Gene Youngblood's keynote lecture, 'Self-Observing Systems: The Video Revolution in America', took place at the Ellicott Complex at the North Campus of SUNY/Buffalo. Author of the influential 1970 book, *Expanded Cinema*, he eschewed his earlier focus on specific tools to speak more generally about the need for a large-scale restructuring of the mass communications system from centralized one-way distribution to consumers to a specialized two-way, point-to-point, user-controlled feedback system.

This was to be a revolution through the total inversion of the structural and functional organization of the mass media, replacing 'the processing of centralized output with processing of decentralized input as the chief characteristic of the medium' (Youngblood 1977), giving the public access to information specified by the user through channels controlled individually. There was at present no mass communication, only mass distribution. Its reorganization would constitute the first true communications revolution in history – the invention of movable type excepted – rather than a mere series of technological innovations.

Youngblood tied the effect of the mass media to human evolution. Influenced by the writings of the Chilean biologist and cognitive scientist Humberto Maturana, his underlying premise was that there appeared to be no grounds for human unity in social modes of thinking and organization other than the desire for it, and that the mass media, by offering only one monolithic version of reality, served to attenuate the passive receiver's ability to imagine others. The interactions of two or more people engaged in communication are requisite to formulating models of what can and should be new structures for talking, thinking, and actualizing their own internal representations of reality. The goal of Youngblood's proposed model was to be achieved with newly (or soon to be) available technologies.[3]

Ken Knowlton offered something like a second keynote address the following morning in 'Technological Art: Some Problems'. Knowlton worked for many years at Bell Laboratories in Murray Hill, New Jersey, designing and implementing software for computer graphics and films, most notably the programming language BEFLIX (Bell Flicks), the first bitmap movie-making system. With his collaboration, these were made available to artists like Stan Vanderbeek, Lillian Schwartz and Laurie Spiegel. Knowlton is also a recognized artist in his own right.[4]

Knowlton used the occasion of 'Design/Electronic Arts' to raise three questions about art as it was coming to pass in the always-morphing environment of new technologies, and the manner in which these technologies should be employed and presented to novices. The first issue was the degree to which the artist should explain the ways in which a work was created, that is, the equipment and processes themselves and how they were used, rather than the particular aesthetic decisions made from a landscape of possibilities. The second was whether collaboration was desirable or more a matter of necessity, given the types of technical knowledge required that might otherwise keep artists at bay. And finally, there was the question, as he put it, of whether one should take 'big steps or little steps in defining new media' (Knowlton 1977), the first ostensibly creating a baffling and alienating hurdle for those who encountered challenging works, and the last representing a set of incremental transformations that would connect new forms with those that came before.

Knowlton's ruminations can be summed up as follows: how to use technology without letting it interfere with the perception of the role of the artist who employs it; and how to initiate a new generation of viewers/auditors into these radical forms in some measured manner.

George Chaikin of the Brain Research Laboratory at the New York Medical College, in a presentation entitled 'The Geometry of Consciousness', proposed a model of visual encoding based on logarithmic spirals – such as one might find in the nautilus shell and the architectural design of the Parthenon – rather than linear logarithms. He speculated that this new model would more closely resemble the way the eye functions, a fact, he said, recently verified by scientific studies. A machine retina could be made not only to be more fully representative of human optical systems, but also to allow for more rapid computation of electronic images, and he had simulated an elementary version of a hypothetical camera that would scan in this manner. This intriguing concept had resonance beyond the manufacture of concrete instrumentation for art to that of the nature of human perception itself.

Music

The development of electronic audio tools long predates that of visual systems. The first is generally considered to be Dr Thaddeus Cahill's Telharmonium (1897), but it was Lee DeForest's invention of the first vacuum tube in 1906 – which would also lead to radio broadcasting and early computers – that spurred a great deal of instrument making for audio signal processing in the 1920s. The most famous of these is the Theremin (1919) by the Russian inventor Leo Theremin. Many prominent composers like Edgar Varèse and Darius Milhaud employed this and other early analog instruments. In the early 1960s, Don Buchla introduced the Buchla Box, and Robert Moog created his commercial audio synthesizer, the Moog. Both were modular in design and used the transformation of tone-generated signals in an additive and subtractive manner to alter pitch, tone and other aspects, and could be played live. These early analog synthesizers were models to a large extent for those who later embarked on building video synthesizers.

In the meantime, digital computer music was also making headway, with systems being developed by Max Mathews at Bell Labs and others in the 1950s, and explored at the Electronic Music Studio at the University of Illinois at Urbana-Champaign founded by Lejaren Hiller, and the Columbia-Princeton Electronic Music Center by Otto Luening, Vladimir Ussachevsky, Milton Babbitt and Roger Sessions. Like many digital video explorations two decades later, these were reliant on large mainframe systems, and most primarily focused on the use of the computer as a compositional rather than performance tool.

One afternoon of the conference was devoted to the most recent developments in computer music by Laurie Spiegel, Joel Chadabe and Joseph Olive.

Laurie Spiegel, a highly respected composer/performer as well as designer of tools for computer music,[5] also worked in the mid-1970s at Bell Laboratories. There, she began to experiment with computer video graphics, using a Rand Tablet and FORTRAN IV software program she'd written for drawing directly onto a display screen, and VAMPIRE (Video and Music Program for Interactive Realtime Explorations). Her talk, 'Designing in Time: Common Principles in Visuals and Sounds', covered the ways in which she interfaced visual composing with the time-structuring software she was already using for music compositions, and the similarities and differences of working in both of these forms.

Joel Chadabe, a Professor of Music and founder of the electronic music studio at the State University of New York at Albany, presented 'A Description of the Landscape Within Which Computer Music Systems Are Designed', which was later published in the *Computer Music Journal*. Chadabe had been a pioneer in the development of real-time interactive systems, attempting to bring the level of feedback of the visual arts to music, and here he spoke broadly about the organizational design of systems that incorporated the digital computer as an important element, either as performer or composer-performer. He used the term 'memory automation' to signify the computer's capacity for storing a score and playing it back for the composer who could then interactively 'conduct' the work. He distinguished this from 'process automation', whereby computer-generated data was based on rules specified by the composer, who was then free to make selections from the outputted sequences. But it was toward the simultaneous composition and perception that Chadabe had concentrated his greatest focus, a procedure that allowed for constant evaluation and flexible redefinition of rhythm, speed, articulation and other details of the music as they were being generated.

Joseph Olive, also a researcher at Bell Laboratories, spoke on 'Applications of Speech Synthesis to Music'. After a brief overview of centuries-old endeavors to simulate human speech by artificial means, he explained the use of spectrographic analysis in attempts to emulate resonances and the like by means of computer software that dominated his text-to-speech synthesis research. His interactive programming was able to alter the time axis, the pitch contour and the sound spectrum. Examples of all of these experiments were interspersed throughout, but the talk culminated with an excerpt from his own opera, *MA-RI-IA-A*, its humorous libretto an enactment of a scientist teaching his machine to talk.

3D motion graphics

In 1977, 3D motion graphics systems, now so commonplace, were in their infancy, with the largest developmental support coming from investors outside the art world. The ability to create only 2D imagery for the video screen was now being extended to manifest representations of synthetic worlds along three axes, a capacity that would have a profound effect not only in industrial, scientific and entertainment endeavors, but those of artists as well.

Therefore, one morning session was dedicated to three-dimensional computer animation systems, or what Charles Csuri referred to as 'visible surface calculation', for motion graphics. The focus was simulation of naturalistic objects and environments for a variety of applications in science, industry and art, and the presenters were James H. Clark, Charles Csuri and Larry Elin.

James H. Clark of the Information Sciences Department at the University of California at Santa Cruz ('3D Digitizing and Motion Description') had been involved in tri-dimensional interactive systems, computer-aided design and geometric modeling since he was a graduate student at the University of Utah. There, he had worked with the legendary Ivan Sutherland, who, among his many other accomplishments, created the first Virtual Reality (VR) system that simulated a 3D environment through which the user could navigate. In 1977, Clark described a system he was developing for NASA that would generate a mathematical model of a wind tunnel to analyze potential flaws in the design of aircraft in real time, and had undertaken the creation of a more sophisticated VR head-mounted display.

Charles Csuri was a Professor of Art and the founder of the Computer Graphics Research Group at Ohio State University. He came from a traditional painting and drawing background, but discovered that programming line plotters could offer a vast array of variations on a theme – again many more than he could have envisioned – through both calculated parameters and random-generated sequences, and created the 1967 film *Hummingbird* based upon them.

At Ohio State University, he was developing the ANIMA II system with artists in mind. Essential design aspects were the use of a script language, or what he called 'scene directives', the use of joysticks for fine-tuning the positioning of an object, and a system of orientation notations that appeared on-screen to assist the user in how to proceed. Defining figures was achieved by combining simple geometric objects that Csuri likened to sculpting, which were then subjected to transformation routines to enact scaling, movement and rotation. Numerous other routines refined this shaping of entities, including the warping and bending of specific planes to enhance naturalistic representation. For the most part, real-time interaction was again considered the key to the successful results of ANIMA II.[6]

Larry Elin came from neither the academic nor the art world, but was, rather, a filmmaker/producer for the Mathematical Applications Group, Inc. (MAGI) in Elmsford, New York – one of the breakout companies for the simulation of three-dimensional solid objects in motion for film and television, using its digital animation process, Synthavision. Much of Elin's work in 1977, as exemplified by his sample reels, was created to assist in the design and application of industrial, military and scientific systems, but there was also a smattering of entertainment-related excerpts that would prove to be precursors for what several years later would be MAGI's most visible project – the production of sequences for the 1981 feature film, *TRON* (dir. Steven Lisberger).

His talk defined the three elements an animator using Synthavision had to address: creating the 3D object shape through a process like Csuri's, here called 'Combinatorial Geometry'; defining an imaginary camera with a location, lens and direction from which to view this synthetic object; and determining a light source by which the scene was to be illuminated. The last of many subsequent stages involved recording each frame of the calculated animation onto film stock. Clearly, this was not a real-time system in any sense, and the results had a 'cartoony' feel that would serve it well when a fellow MAGI employee, Jon Lasseter, cofounded Pixar Studios.

Graphic arts

Aaron Marcus and Sonia Sheridan gave talks on the use of new technologies for the graphic arts – from computer-assisted design to the aesthetic applications of xerographic systems.

Aaron Marcus was a designer and computer and conceptual artist from the School of Architecture and Urban Planning at Princeton University. His work was in 'Visible Languages', a term he used to describe calligraphic and typographical signs, drawings and markings as a 'useful tool for examining systematic aspects of diagrammatic compositions of iconic and symbolic forms' (Marcus 1977). During a period in residence at Bell Labs, he developed a 'prototypical, interactive, computer-assisted page design system' (Marcus 1977), which permitted the flexible electronic arrangement and manipulation of signs and symbols on the surface of a cathode ray tube. Many of the elegant works Marcus projected bore resemblance to concrete poetry, where the relationships of figure to field, geometrical components and other formal properties became aspects of expressive alphabetically-based art.

Most pertinent to the general conceit of the conference were slides from his dynamic series *Cybernetic Landscapes*, virtual realities of a sort that he described as 'poem-drawing environments'. These were highly interactive systems whereby the viewer, by means of joysticks, was able to travel through a dark landscape upon which had been designated signs, symbols, letter forms, objects and other graphic elements rendered as pure-white line drawings.

Sonia Sheridan was the creator of the Generative Systems Program at the University of Illinois at Chicago Circle Campus, a remarkable facility that functioned as a research center for student artists of what she referred to as 'The Missing Media' between video and computers; as a place that wedded earlier methods of art making, such as drawing, photography and textile design, to new electronic technologies; as an interactive force between industry, education and the public; and, perhaps, most importantly, as an attitude and process that extended beyond the use of any specific machine: 'It attempts to reverse the order of motion, from machines remain, to humans remain while machines move' (Sheridan 1975: 6–10).

Sheridan's work focused on Electronic Systems Printout (ESP) using a variety of industrial copying equipment tools such as 3M's Color-in-Color machines, Xerox copiers, and other means of thermographic and electrostatic reproduction. Rather than the more conventional relationship between artist and industrial scientist/technician – where the latter might serve largely as instructor of the machines of the former – Generative Systems often gave back new methods in which these complex pieces of technology could be differently employed: '[I]ndustry has a backlog of 50 years of inventions for which they are looking for new uses, applications and markets. The artist who has access to these inventions is well equipped to dream of alternative inventions and applications' (Sheridan 1975: 6–10). Concomitant with that approach was the recognition that scientists were as creative in their activities as artists, albeit with different aims in mind.

That the systems were interactive and largely in real time was extremely important to this approach as innovative ways to engage them. In this, the thrust of Generative Systems was not unlike the attempts of video and computer artists to provide rich feedback to the creative user.[7]

Design by collaborative teams for computer-based video systems

The portion of the conference most plentifully represented was that of close collaborations by small cadres of artists/engineers to realize hybrid analog/digital video synthesis tools in Chicago, Albany, New York City and Binghamton. This model was, without question, the most productive of any, and that which demonstrated an ideal for design in the electronic arts.

3.1. The Chicago Group – Dan Sandin, Tom DeFanti and Phil Morton

Dan Sandin, Tom DeFanti and Phil Morton created the Electronic Visualization Center, and referred to it as a 'habitat' – a felicitous term signifying an environment in which one could work for indefinite periods of time, continuously learning new ways to yoke the system to her or his own creative ends.

Dan Sandin of the Art Institute of Chicago, then just undertaking the digital version of his analog Image Processor, reviewed the design principles that had gone into it, and these have been described above. One further quote makes yet another salient point:

> I designed the IMAGE PROCESSOR from the point of view of, what does the student have to do to use it? [N]ot from, what can the machine do? [W]ith electronic visualization done properly, the feedback is no longer the *limit*; the limit then becomes how fast your mind can process information; and that is a *much higher limit*, in some cases, especially when processing visual information. (Sandin and Morton 1977)

In Tom DeFanti of the Department of Chemistry at the University of Illinois at Chicago Circle Campus, a formidable computer software designer, Sandin found a kindred spirit with whom to collaborate. DeFanti's GRASS (GRAphics Symbiosis System) became the basis of interfacing with Sandin's analog processor, and DeFanti said that he almost thought of GRASS as an analog system, in that it allowed him and Dan to appropriate real power to accomplish a great deal in short order. He also was a firm believer in high levels of interactivity – to make the system so unsophisticated that it could operate in real time – and in tactility, with large knobs and other paraphernalia. Digital-to-analog conversion made Sandin's analog IP and DeFanti's GRASS software entirely compatible.

Phil Morton, also of the Art Institute of Chicago, was one of the primary users of Sandin's and DeFanti's system, and took it in directions that were distinctly his own. A keen advocate of Citizens' Band radio in particular as a model adaptable to video, he was deeply engaged in the whole topic of telecommunications in general.

Morton forayed into the realm of interactive communications by way of high performance two-way video and audio 'transmission' installations. These were modest (but doable) in scope compared with Youngblood's convulsive model. The typical staging of a Two-Way Communication System consisted of three basic components: (1) an observation area for contemplating the ongoing process and resultant product; (2) two public user terminals, each consisting of a video camera and microphone trained on the other to send information with a monitor and headphones to receive signals from the other site, and tactile devices allowing modification of the focus, angle, intensity and scale of the camera coming from the opposite terminal; and (3) a processing area, where one person (termed the CO-OPERATOR) mixed the camera outputs into a single recordable channel, a second performed on an Arp music synthesizer (the ARPIST), and a third processed the images through Sandin's tool (the IPIST).

3.2. Tom DeWitt and Philip Edelstein: Pantomation

Film and videomaker Tom DeWitt and electronics designer Philip Edelstein from the Electronic Music Studio at SUNY Albany made a joint presentation about a new multipurpose tool they had devised, The Electronic Pantograph, which had applications in music, dance and other temporal arts. Edelstein, with others, was responsible for the basic hardware, The Key Positioner, and the specific application by which the process was designated – pantomime – the particular interest of DeWitt. The Pantograph, DeWitt noted, 'uses the eye of a television camera and the brain of a computer to keep track of moving points in space and writes a record of this movement in the abstract language of a computer-choreographer' (DeWitt 1977). It was a graphic notational system of body movements, in particular, achieved by attaching a yellow spot to the performer that was followed by a camera employing chroma key techniques. The computer component provided additional flexibility through its control language, such as, according to Edelstein, 'keeping an object in a fixed position in a video frame although it was originally moving in the real world'

(Edelstein 1977), as well as retracing the described movement by storage of horizontal and vertical components in computer memory. Excerpts of the first use of the Electronic Pantograph, DeWitt's comic sci-fi pantomime, *Outta Space*, demonstrated this process.

3.3. Lou Katz and William Etra

Lou Katz was working at the College of Physicians and Surgeons at Columbia University making 3D models of scientific objects when he developed an interest in using its large computer to create aesthetic entities as well. As this work developed, he began to consider rules by which to create images of greater symmetry and complexity that could be changed for the purpose of experimentation and further be preplanned and controlled for repeatability.

At this point, he became acquainted with the artist William Etra who, with the designer Steve Rutt, had created one of the earliest video processors, one which operated on very different principles than did other analog synthesizers of the time. Katz took notice of the possibilities of working with video, and he and Etra began to collaborate on developing a computer-controlled video system. This included finding the means to convert terminal displays and feed them into synthesizers and other effects tools. Like Sandin, Katz also began to make voltage and image signals interchangeable, so that one picture element could be used to control another.

Etra was introduced to lightweight portable video equipment as a film student at New York University and, after having been exposed by Walter Wright to Scanimate, a commercial analog synthesizer, he teamed with engineer Steven Rutt to construct the above-mentioned Scan Processor. While working in an electronic music lab, Etra became convinced that computers constituted the next step, and several years later acquired his own microprocessor. One of the attractive uses of the computer, he noted, was as a notational tool for greater control and repeatability, and the viability of new types of peripheral devices.

3.4. Donald McArthur and Walter Wright: A Computer-Based Video Synthesizer

The penultimate session consisted of a two-part presentation by Donald McArthur and Walter Wright about the construction of 'A Computer-Based Video Synthesizer' at the Experimental Television Center (ETC) in Binghamton, New York. McArthur designed the hardware – aka the Spatial and Intensity Digitizer (SAID) – and Walter Wright was responsible for the software for system operation.

McArthur's background was in theoretical physics, and while teaching at the State University of New York at Cortland he became acquainted with the work of the ETC. In 1975, he and Wright initiated the computer-controlled video system, based on the use of a microcomputer interfaced with McArthur's own self-designed digital-to-analog conversion circuitry to provide a NTSC video signal that could be controlled in real time, used for live performance, and recorded on standard videotape recorders.

Rather than paraphrase McArthur's comments at the conference, it is better to quote him at some length describing the functions of the SAID architecture from a paper written in June 1977, functions that bear resemblance to Joel Chadabe's own design concepts:

> There are basically two modes of operation of the system: interactive-compositional mode and automatic-production mode. In the compositional mode, the artist can enter programs and parameters through the keyboard, observe the resulting sequence of images, and then modify parameters through either keyboard or a real-time input and thus build up a data set for a complete piece. The data set, representing all the aesthetic decisions made by the artist, is stored in the computer at each stage of the composition. When the composition is finished the system will operate in the automatic-production mode generating the final video signal in real time with no intervention by the artist. The artist may also choose to use a combination of these two modes in an interactive performance or to allow an audience to interact with the system operating automatically. The system is structured so that all of these variations can be accommodated by appropriate programming. (McArthur 1977)

McArthur spoke further of some of the design principles that went into the construction of this complex system consonant with those expressed by others, including that it be based on a general-purpose machine operating in real time – modular, hybrid and portable.

Walter Wright's talk served as complement to McArthur's presentation, approaching software programming to establish his working relationship with the microcomputer and video synthesizer. Being an artist foremost, he began with the end product – the image – and basic rules of composition that needed to be considered. Wright also extended feedback beyond the interaction between the artist and image to include that between synthesizer and computer, achieved by factors like predictable unpredictability, conditional (if/then) branching, and setting the computer to reprogram itself under certain conditions. This creative programming made the Computer-Based Video Synthesizer a very versatile tool with the potential for all manner of production, presentation and storage.

Evening events: presentations, a mini-retrospective and a concert

After long days of fervent interchange, evenings were given over to more palliative presentations. Composer Robert Ashley showed excerpts from his recently completed series of unconventional video portraits of his peers entitled *Music With Roots in the Aether* (1976). John Whitney, the dean of new technologies as inventor and artist employing systems for abstract animation, was scheduled to give a talk entitled 'From Raster Scan to Slit-Scan'. In its stead, he concentrated on his most recent work, the stunning *Arabesque* (1975), which was the first film that he made using digital rather than self-designed analog equipment, and is a complex seven-section fluid weave of intricate curvilinear patterns composed of a 360-point array in constant movement. Whitney offered a close reading on an analyzer film projector of the harmonic principles, akin to those found in music, that underlie its structure. This represented his most current concern with the periodic organization of perceptual experience, and a fully illustrated essay in the same vein, along with a broader look at his career development, can be found in his indispensable book, *Digital Harmony* (1980).

Immediately following Whitney, Stan Vanderbeek presented a mini-retrospective of his work over the previous two decades, beginning with camera-less and cut-out animation films that began his career, through his computer work with Ken Knowlton creating a series of *Poemfields* at Bell Laboratories, and beyond. He also discussed a number of other innovative projects he had undertaken, including projection of film images onto steam, and eight-hour night-long *Cine-Dreams* staged in a planetarium, where viewers were encouraged to drift in and out of sleep to a barrage of images projected on the dome, and later report their oneiric experiences by telephone. Vanderbeek specifically referred to the planetarium as a 'tool', and made use of its unique imaging systems in addition to his own work.[8]

On the final evening, the members of the Sonic Arts Union – Ashley, David Behrman, Alvin Lucier and Gordon Mumma – each performed one of their musical compositions in the Grand Court Lobby of the nearby Ellicott Square Building. Their relevance to the concerns of the conference is best indicated by the title of Behrman's piece, *Music for Cornet, Micro-Computer and Homemade Electronics*, which credits as performers Jim Horton on computer programming and Kim One, his microcomputer, on harmonic changes.

Woody Vasulka: a summary talk

It fell to artist Woody Vasulka, a professor at the Center for Media Study, SUNY Buffalo, to provide final observations and an open exchange with the audience on the ideas expounded throughout the conference. He prefaced this discussion with his own conception of what he called 'The Territory, The Subject, and the Transmission of Knowledge'.

His notion of 'The Territory' was that of placing development of new artists' tools in the context of a particular era and certain contemporaneous technological innovations. In looking at 'The Subject', he proposed that new concepts arose in electronic imaging that differed substantially from those of photography and cinema. This subject was that which comprised the electronic image as a time-energy construct, such as waveforms and control signals. Vasulka's own personal proclivity was apparent here, as an investigation of the primary elements of the video image (or, as Walter Wright had put it, finding out 'what makes TV tick') had dominated his work from the outset, and he had often referred to his own activities as dialogues with tools, an attempt to understand them and what they revealed about the medium.

Regarding 'The Transmission of Knowledge', he strongly promoted greater dialogue between individuals in transferring of information on hardware and software. Not only was there secrecy in the military-industrial complex, as one example, but also, in what seemed a misplaced implication, between members of the 'tribe' – his term for the very sorts of participants involved in this four-day conference – who seemed to have been engaged in nothing so much as information interchange.

After his quasi-polemic, Vasulka opened the floor for an extended period of discussion in which a plenum of subjects were touched upon, among them: that an artist reflects who he or she is irrespective of the system employed; whether too much emphasis has been placed on innovation and that the idea of being first diminished the importance of being best; and whether the system and program, as well as the resultant image/sound, were, in fact, a work of art.

4. The subsequent years, 1977–2010: some thoughts

As noted above, there were some pragmatic reasons to distinguish between artist and engineer in the electronic arts in the late 1960s and 1970s – the complex technical knowledge necessary to write software for institutionally based computers, and access to artists from those working within these facilities being the two most prominent. But a number of these engineers saw themselves as artists as well, and established makers were already conversant with software design and how it could be applied to their desired ends – this too was part of their creative process. The question raised by some conference attendees as to whether the design of hardware and software systems could themselves be considered an art form clearly depended on how broadly or narrowly one wished to define 'art'. Perhaps this convergence of both roles is best summed up in the following quote by Andy Hertzfield of MIT's Media Lab:

> It's the only job I can think of where I get to be both an engineer and an artist. There's an incredible, rigorous technical element to it, which I like because you have to do very precise thinking. On the other hand, it has a wildly creative side where the boundaries of imagination are the only real limitation. [...] It takes incredible concentration and mind space. (Brand 1987: 58)

In the more than three decades since, an entirely new information techno-culture has evolved that has made the kinds of capabilities envisioned in 1977 so pervasive as to be commonplace to a new generation weaned on digital technology – including emerging artists.

It would be an exaggeration to say that analog technology has become obsolete – but only by a fraction. Although there are still pockets of artists making use of analog tools here and there, in the larger context conventional television has been broadcast solely digitally throughout the United States since the middle of 2009. That DTV may in fact cause some degradation in image quality and color distortion is largely overlooked – much as had been the case with the transfer of analog to digital audio, except among true aficionados.

Personal computing devices of many degrees of portability and versatility have, of course, become ubiquitous – perhaps even more so than TV sets themselves. Software packages for computer graphics are often part and parcel with these devices and otherwise stock shelves. Some of the techniques artists were struggling to obtain in 1977 can be readily learned in coursework at art schools and other institutions. And, needless to say, 3D computer animation of all sorts is inescapable.

And then there is the rise of the Internet – Sir Tim Berners-Lee's World Wide Web – the apotheosis of a true communications revolution as defined by Gene Youngblood. No one in 1977 could have envisioned this phenomenon, but in the words of the great performance artist Yogi Berra: 'It's tough making predictions, especially about the future'.[9]

The new form of Internet art (or net art)[10] developed in the mid-1990s, but the extent to which this phenomenon has been of great moment except to the relatively few cognoscenti is difficult to gauge, and what sorts of work it will produce in the future remain to be seen. A notable contribution has been made, surprisingly, by the website YouTube, which has created a new outlet for the distribution of video art. Virtually every artist who appeared at the conference has works deposited there for viewing in an easy-to-access format.

This is not the place, nor is it my intention to make any sweeping comments about the current state of electronic art. Certainly a great deal of video appears in galleries and museums, and has become wholly integrated into mixed-media installations and performances of all sorts, as a result, in part, of high-quality digital cameras and projection systems. The art world has accommodated and even embraced video as it had earlier ghettoized it.

At first consideration, it seems that less of the kinds of expression we once identified as electronic art – e.g., video synthesis and image processing – has made the transition into the present. But I would qualify this by the reflection that work of this nature was always a subgenre of video in general; many, if not the preponderance, of early tapes and installations evolved around the conceptual and minimalist art movements that often employed apparatus little more sophisticated than ½" monochrome recording equipment. The apparent dearth of electronic video may also have much to do with a general trend away from abstraction and the fact that the novelty of video synthesis has waned over time.

Finally, has anything been lost in the retirement of analog image-processing tools? From my own perspective, gone is some degree of idiosyncratic charm or personality that we associate with handicraft and folk art, by way of machines that reflected in their design and use the human touch that may be missing from many of today's far more sophisticated systems. But the trade-off so substantially favors current technology in other respects as to make this sentiment largely irrelevant. Unearthed artifacts of the Stone Age tell us much about the nature of the progress of civilization, but hardly inspire nostalgia for a lost innocence and quality of life prior to the development of new industrial and post-industrial tools. To the extent that this essay has been an archaeological enterprise of a sort, the conference participants it documents were continuously striving for the new tools that would constitute the next step in transforming aesthetic culture.

References

Brand, Stewart (1987), *The Media Lab: Inventing the Future at M.I.T*, New York: Penguin Books.

Burroughs, William S. (1971/6), *Electronic Revolution*, Germany: Bresche Publikationen, pp. 21–31.
——— (1983), *The Place of Dead Roads*, New York: Holt, Rinehart and Winston, pp. 215–17.
DeWitt, Tom (1977), 'Pantomation', paper distributed at 'Design/Electronic Arts' conference, March 10–13, Buffalo, NY.
Edelstein, Phillip (1977), 'The Key Positioner: A Device for Implementing Pantomation', paper distributed at *Design/Electronic Arts* conference, March 10–13, Buffalo, NY.
Knowlton, Ken (1977), 'Technological Art: Some Problems'. Quotes from audiotape recordings of conference.
Marcus, Aaron (1977), 'Visible Languages: Towards Diagrammatic Environmental Communications'. Quotes from audiotape recordings of conference.
McArthur, Donald, Walter Wright, and Richard Brewster. (1977), 'A Computer-Based Video Synthesizer System' [unpublished paper].
Sandin, Dan and Morton, Phil (1977), 'Research Report on Public Media', paper distributed at *Design/Electronic Arts* conference, March 10–13, Buffalo, NY.
Shannon, Claude F. (1962), 'Computers and Automation – Progress and Promise in the Twentieth Century', in S.W. Higginbotham (ed.) (1963), *Man, Science, Learning, and Education: The Semicentennial Lectures at Rice University*, Houston, TX: Rice University, p. 210.
Sheridan, Sonia (1975), 'Generative Systems: Six Years Later', *Afterimage*, 2: 9, pp. 6–10.
Whitney, John (1980), *Digital Harmony: On the Complementarity of Music and Visual Art*, Peterborough, NH: Byte Books/McGraw Hill.
Youngblood, Gene (1977), 'Self-Observing Systems: The Video Revolution in America'. Quotes from audio tape recordings of conference.

Notes

1. There had been a longer tradition of the development of electronic music synthesizers. Those most recent and influential had been Don Buchla's Buchla Box (1963) and the first Moog Synthesizer (1964), both of which provided models for the design of the early video synthesizers. An excellent source for information about early visual and sound tools is *Eigenwelt der Apparate-Welt: Pioneers of Electronic Art* from 'Ars Electronica', 1992, an exhibition curated by Woody and Steina Vasulka.
2. It is interesting to compare Sandin's list of desired design principles with those articulated for the industrial computer put forward in 1962 by Claude F. Shannon in his lecture, 'Computers and Automation – Progress and Promise in the Twentieth Century': 'The next generation of computers will be faster, smaller, with greater flexibility and memory capacity, and more reliable. We expect the programming to progress so that it becomes easier to communicate with computers using our own language or something close to it' (Shannon 1962).
3. Another, albeit unusual, reflection of technological innovation and its suppression, as well as biological evolution in this context, can be found in William S. Burroughs's hallucinatory novel, *The Place of Dead Roads* (1983: 215–17). See also *Electronic Revolution* (Burroughs 1971/76: 21–31).
4. See http://www.kenknowlton.com/.
5. See Laurie Spiegel's website, http://retiary.org//ls for extensive information about her software design for digital arts, as well as her music and other activities.
6. In 1977, Csuri's system had been in development for only a few years and was still developing new features. A more comprehensive expression of the artist's ideas, circa 1975, is available at http://www.atariarchives.org/artist/sec25.php.
7. The specific applications that Sheridan and her students employed are too numerous to list here, but an archive of her work has been preserved at The Daniel Langlois Foundation for Art, Science and Technology (La fondation Langlois) in Quebec, Montreal.
8. A somewhat fuller description of Vanderbeek's first *Cine-Dreams* event at the Strasenburgh Planetarium in Rochester, NY in 1972, which I attended, was published in *WNET News*, 1: 3, and it quotes Stan as saying that dreams are 'a rehearsal for the future. Cronkite, instead of reporting the weather, may one day report that 3,000 people dreamed of earthquakes last night'.
9. Claude Shannon, again, spoke in more concrete terms about the pitfalls of predicting future developments, also in the Rice lecture: 'With the explosive growth of the last two decades in computer technology, one may ask well what lies ahead. The role of the prophet is not an easy one. In the first place, we are inclined to extrapolate into the future in a straight line, whereas science and technology tend to grow at an exponential rate. Thus our prophecies, more often than not, are far too conservative. In the second place, we prophesy from past trends. We cannot see the great and sudden mutations in scientific evolution. Thus we find ourselves predicting faster railroad cars and overlook the possibility of airplanes as being too fanciful' (Shannon 1962).
10. 'Net art' comes in a variety of forms with a broad range of intentions, encompassing everything from e-mail projects and websites to innovative Internet software and participatory networked art enterprises among users. It is an international movement that had its origins in Eastern Europe and is largely interactive. For an excellent history of the phenomenon, see: R. Greene (2004), *Internet Art*, London: Thames & Hudson.

Instruments, Apparel, Apparatus: An Essay of Definitions

Jean Gagnon

One of the many developments of media arts in the last decade is a heightened awareness of the role that machines, apparatuses and devices play in our lives and in the arts. The result is an emerging, complex and risky field of research slowly taking shape to chart this new focus on tools and instruments, or the apparel and apparatuses related to the production and reception of art. Tools and instruments have had long histories associated with material cultures, musical and scientific instruments. Despite the fact that technologies have had a tremendous impact on the arts since the nineteenth century, the question of tools and instruments, the means by which artists shape materials (*l'outillage de l'artiste*), has remained little studied outside of the field of music.

This essay is an attempt to describe what lies at the heart of the instrumentation of art practices as an 'instrumental relation'. To date, terms used to refer to the artist's instrument or about works of art resulting from the use of instruments have been limited to terms, such as 'machine', 'device' or 'apparatus', that remain mostly undefined and are often used interchangeably. Here, borrowing from music where types of works are commonly categorized as 'instrumental music', I want to refer specifically to 'instrumental art' which I use to describe artworks produced by the use of electronic instruments that permit the generation of images, sounds or audiovisual forms in real-time live performance. Before exploring aspects of 'instrumental art', I will attempt first to clarify the terms and definitions of the discussion.

In the following, it is useful to note that I deploy musical instruments as the model to establish a phenomenology of the instrumental relation and playing in the production of audiovisual art forms. However, to arrive at a more accurate description of this relation, I also take into consideration the role and nature of instruments in the field of techno-science, where they have long played a dominant role. Today's electronic and digital instruments share similarities whether they are used by scientists or by artists: they facilitate visualization, and pick up and emit signals, sounds or lights. But whether they belong to music or science, what are the common traits of instruments in general? Once these commonalities are established, the discussion attempts to distinguish specific characteristics of the 'instrument' in contrast to those of the 'apparatus'. The analysis then continues by making further distinctions between these terms and the notion of 'apparel'. Apparel, derived from *appareil* in French, and to which it corresponds, may strike some as archaic English; however, this essay proposes that a reconsideration of the notion may greatly contribute to a more accurate description of some of the more subtle uses of technological devices in the arts.

What is an instrument?

Where does the word instrument come from? Etymologically, the noun is borrowed from the Latin word *instrumentum*, meaning 'furniture' and 'implement'. As a verb, it is also related to *instruere*, 'to arrange' or 'to instruct'. In English, the network of relations between the meanings of the words 'instrument' and 'implement' as nouns and verbs is vast compared to their original derivations. In Latin, the noun *implementum* refers to the action of filling up, while *instrumentum* refers to the means whereby something is achieved, performed or furthered. The implement is therefore the device used in the performance of a task. As verbs, 'to instrument' means to address a legal instrument, or to score a musical performance, to orchestrate; while 'to implement' is to carry out or accomplish something, to provide instruments or the means of expression for something.

Encompassing all these meanings, the word 'instrument' in English thus refers at the same time to both the executive and operational technical dimensions, and its cognitive and cultural aspects, in order to instruct us about something. Put another way, the instrument is understood not only as that which allows us to play and produce sounds but also as a legal function of mediation.

Tools and instruments

Given that tools and instruments are generally conflated terms, refining an understanding of their similarities and specificities will help to better grasp both terms. Broadly speaking, tools and instruments are considered as body extensions, that is, the exteriorization of human capacities operating through the amplification of perceptual functions. The microscope or the telescope are such examples.

Both tools and instruments embody the power of humans to anticipate and imagine. To imagine is to create a future; to anticipate is to preview what has not yet happened. Thus to be human is to be in time. These two human faculties of imagination and anticipation do not float in the immaterial sphere of the 'spirit'. Rather, they manifest through their incarnation and exteriorization and by the actualization of these capabilities.[1] It is therefore important to keep in mind that tools and instruments allow for modes of embodying the power to anticipate and create the future. My insistence on this temporal dimension is to underscore the fact that media arts and new media are fundamentally time-based.

In his philosophy of 'technology' (which he referred to as 'technique'), Gilbert Simondon defines the tool as a technical object extending or gearing up the body to accomplish a gesture, and the instrument as a technical object that enables the body to extend and adapt to obtain a better perception (Simondon 1989: 114).[2] In short, Simondon's distinction is to insist that individual humans seek to enhance their knowledge by using instruments. But Simondon also advocated for a technical culture independent from labor: 'This relation of the human individual to the technical individual is the most delicate to form. It implies a technical culture which introduces the capacity of attitudes different from labor or action' (Simondon 1989: 119). According to him, because tools, instruments and machines are used by humans to achieve practical aims in often unnoticeable ways, as long as they fulfill their duty, developing a technical culture would enable a consideration of technical objects for their own sake and as technical beings. Here I suggest that play might also be an attitude towards technical objects that is different from labor and action.[3]

He continues by affirming that what work and action have in common is the predominance of the finality over causality, while in auto-regulated systems, any causality has a sense of finality, and any finality a sense of causality. Cybernetic systems are auto-regulated. But they are also closed circuits, as dynamic as they might be; they are systems relying on redundancy and, as such, are counter to the openness of imagination. Simondon's solution to escape imprisonment in cybernetic feedback loops resides in the notion of invention. Invention, like creation in art, despite several differences, finds its fuel in the anticipatory nature of human thinking, which delocalizes the present into the future, and in an 'indeterminacy margin' (*marge d'indétermination*) that Simondon posits at the heart of technological evolution. He discusses it in relation to the question of the evolution of technology, and it is relating to notions of forms and information. The form of a signal permits us to distinguish information from noise. An instrument – a machine in Simondon's text – is made out of forms, and through these forms information can occur as an event outside of the predictable. Thus information is not a form; it is the 'variability of forms'.

The indeterminacy margin is therefore constituted in instruments as critical phases in their functioning. The instrument that can receive information is the one that can localize temporally its indeterminacy at critical moments full of possibilities. This structure, according to Simondon, is that of the 'decision' and 'relay' (Simondon 1989: 141). For instance, digital instruments, such as those used in music or those working from body signals and movement sensors, show this indeterminacy at play. This temporal structure of decision and relay is what allows the player to intervene despite the fact that, at times, the system can also take some of it in charge, and it is at the heart of the functioning of electronic instruments at the moment when information comes into play. (See Color Plate 40.)

Musical instruments

Where do musical instruments come from? What has been the moving force behind their invention? Are they a necessary service and pleasure for humans? Among the ethnologists, musicologists and historians who have long sought answers to these questions, André Schaeffner's position offers a promising point of departure: the origin of musical instruments in human societies, according to Schaeffner, is in what unites 'language and singing, dance and instruments' – that of the human body (Schaeffner 1994). The human body's first impetus is to make noise and to shape instruments to respond and correspond to itself and its postural and gestural capacities. While language, singing and dance are in the singular, 'instruments' for Schaeffner is in the plural, as if they could only be conceived in their multiplicity and variability of materials and morphologies. The very multiplicity of instruments testifies to their 'incarnated nature'. They are embodiment linked to human gestures and postures. One could even say that instrumentalists and dancers share this gestural and postural performative relation with music.

A gesture can be seen as dynamic, whereas a posture is often considered static; dance would certainly deny that, showing postures that are always in equilibrium, on the brink of losing mastery. Mastery, here, plays a role as great as in playing musical instruments: both involve the body of the performer although in different ways and manners; both are concerned with 'pleasure through mastery'. But an instrumental gesture has some more features: 'it is applied to a concrete and material object with which there is physical interaction' (Miranda and Wanderley 2006: 10), and the forms and dynamics of that interaction can be mastered by the subject.

According to Schaeffner, one of the most significant aspects of music is its perpetual power to limit tonal sources through the use of a few privileged materials, fixing their resonance to specific degrees of intensity. Concerning instruments, it is only habits and local conventions that establish these limits. It is through this effect, which is also a selective process, that certain sounds emerge and manifest, while others are suppressed or relegated to the margins or simply disappear. The sonic world is organized according to certain materials and timbres that reduce other sounds to silence. This explains the variability of musical systems and tonal canons of instrumental music in different cultures. Shaeffner's point is similar to that made by Jacques Attali in his renowned essay 'Noises' on the political economy of music, in which he argues that music is a regulator of noise in society in order to preserve a certain social order. The powers that govern any society impose codes to analyze, mark, restrain, suppress, train and channel the noises of bodies, language, tools and objects. Schaeffner also reminds us of music's relationships with the rhythms of work, with toys and games (Schaeffner 1994: 108), and with magic (Schaeffner 1994: 117): 'It is not the aim of these pages to grab the part play has in music, but rather the part of sonic material (sound) in play' (Schaeffner 1994: 108). In the same spirit, I would like to suggest that my intention here is not to impose the 'playful' (*ludique*, in French) as a category of new media – this has been done by others – but rather to consider instruments (such as toys and games) as a determinant of 'instrumental playing'. Instrumental playing as it participates in the structure of the playful relation is a central notion, as it qualifies the relation by pointing to its performative aspect, with a particular embodiment for the instrumentalist for whom the play's immediacy is apprehended through the mediation of his or her knowledge and mastery of the instrument. It is worth mentioning with another author, Hugues Dufourt, that one of the founding moments of Western music and culture took place a long time ago when the Pythagorean used an instrument called a *canon*, which produced both the sound (a vibrating string) and its measurement (Dufourt

1991: 247). Thanks to this instrument, Pythagoras and his disciples could imagine the isomorphism of sound and number; with them, Greek thought founded musical science on mathematical concepts. The Pythagorean harmony would not have been possible without geometry, which made the visual transposition so determinant (Dufourt 1991: 246). The Greeks did not go further in this vein and were satisfied with this conception of proportional isomorphism between spatial relations and numbered relations, which allowed them to study relations of consonance (Dufourt 1991: 250). It is through the use of an instrument that they were able to establish iconic visualization of sounds in the form of geometric and numbered representations.

This has opened a long dialectic in Western history, articulating the relative values and powers of a visual and an aural paradigm[4] that, I suggest, underwent a shift more recently when electronic audio-video instruments, and later digital instruments, appeared to allow the visual artist to gain the immediacy of instrumental playing while giving musicians and composers a live visual system with which to create music and sounds.

In his 1991 article on scientific instrumentation, philosopher Don Ihde noted the similarity of electronic and digital instruments in music and art and those of science: 'The emergence of the sound studio which bears a direct parallelism with a science laboratory, provides the possibility for further manipulation and transformation of sounds' (Ihde 1991: 22). Returning to ideas developed previously about scientific instruments, he noticed in musical instruments what he called their 'multi-stability' in reference to how their use is transformed by the context. For instance, the bow that is used to hunt can become a stringed instrument by adding a resonator, a function that can sometimes be accomplished by the mouth (Ihde 2007). Therefore, human action is embodied in multi-stable instrumentation which serves both survival and artistic musical playing.

Scientific instruments

Informed by phenomenology and Merleau-Ponty's views on intentionality, Ihde sees scientific instruments as embodiments of knowledge, suggesting that they are 'in use and in relation to users', that they cannot be conceived 'apart from their context of involvements and referentialities' (Ihde 1991: 73). The technical object cannot be conceived outside of the environment and the human subjects within it with which it interacts. This aspect of how humans and instruments relate in and to the world is essential to understanding technical objects. To use Simondon's terms, the technical object is a mixed milieu that is both technical and geographical, and in which individuals intervene. For Simondon, humans are the true mediators or mediating operators between the technical and geographic worlds, and between technical elements, technical individuals and technical ensembles.

While instruments can be seen as 'extensions' of the body, or as enhancements of human perception, according to Ihde, there exists a second order of relations for instruments, 'a second group of relations [that] does not extend or mimic sensory-bodily capacities but, rather, linguistic and interpretive capacities' (Ihde 1991: 75). In addition to the more transparent first order (experienced-through), there is a second order composed of degrees of opacity (experienced-with) as if instruments were accompaniments that enhance and refine our perceptions of the world rather than a means that disappears when used. The more complex the instrument, the more they require a second order of relation with the user, which Idhe describes as a 'hermeneutic relation', where 'the technology is not experienced-through as experienced-with' (Ihde 1991: 75).

Ihde sheds light on the amplification-reduction structure of instruments that shapes their evolution, and creates a friction between the two structural terms, a friction he calls the 'opacity' of instruments. According to the degree of opacity of a given instrument, it requires a more or less sophisticated hermeneutic knowledge as to how to read and interpret its results. This friction occurs along a graduating line that goes from the ideal and desired state of total transparency, when the instrument mediates and enhances human senses (microscope, telescope), to the other extreme of maximum opacity that requires hermeneutic procedures, learning and knowledge (spectrographic imagery). Certainly, notes the author, this dream of the purely transparent device points to a developmental *telos* for technological advances. The amplification-reduction structure in data-gathering instruments is uncovered as intertwined within the context and the situation in which it occurs with the intentionality of human subjects.

Scientific and musical instruments

Now let us ask ourselves what is the difference whether one strives for increasing one's knowledge and power through the use of instruments or, instead, one's pleasure and imagination through instrumental playing, as in music and art? While the scientific instrument is ideally meant to be as transparent as possible, the musical instrument does not wish for transparency. Timbres of musical instruments, among other things, favor their opacity. It will be through a long and often painstaking practice that instrumentalists will eventually reach mastery of their instruments, therefore attaining transparency by mastering the opacity of an instrument. Don Ihde arrives at the same conclusion when he writes: 'But as skill is acquired, the flute is "mastered" in that it withdraws or becomes more and more transparent [...]' (Ihde 2007: 11).

To schematize this distinction between the scientific and the musical instruments, I may propose the following equation (Figure 1):

> Now, this equation applied to the musical instrument presents a somewhat reversed situation: instruments, in diverse degrees, are opaque and only a few of them can be played without real training. One of the most difficult, the violin, can be very annoying played by an untrained person. So, to learn an instrument requires knowledge acquisition and active practice in order to master it and play some music. This mastery

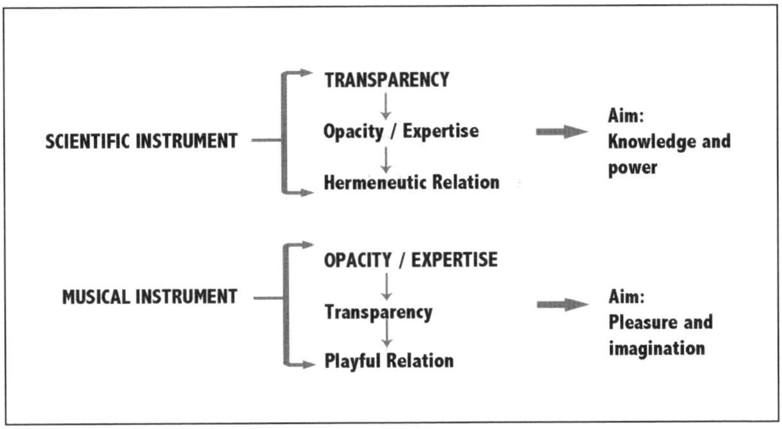

Figure 1. Difference between scientific and musical instruments.

of the instrument necessitates that of the body at play through a particular dexterity and posture. From this embodiment with instruments springs the playful.

Don Ihde proposes an 'instrumental realism'. We affirm here an instrumental playfulness (*ludique*). Two types of instrument are here contrasted: those that strive for transparency and end up in this hermeneutic relation where the instrument's rendering of reality is to be interpreted; and those that we encounter first as opaque because of timbre, in the case of musical instruments, or of other marks of their materiality, which through a playful relation, a mix of mastering the instrument and having it lead the way, attain transparency in performance. Playing musical instruments and the music played on them can have aesthetic or semiotic values and symbolic resonances.

Both the hermeneutic and the playful relations suppose mastery and knowledge but for different aims: acquisition of data and knowledge in one case and aesthetic pleasure in the other. The two types of instruments share the performative, effective, implemental effect of modifying the state and scale of our perception threshold and of our level and quality of presence in the world.

A musical instrument is most often a technical object. It has something of an individuality, even if it belongs to one class of instruments or another. It possesses a particular morphology and timbre constitutive of its opacity. While most traditional musical instruments are passive, waiting for the player's intervention, when we consider electronic instruments in music we see that some are semi-autonomous – they will perform certain operations that will be fed back to the instrumentalist. The instrument can therefore be conceived as a 'quasi-other', a quality that instruments share with games. In art, though, the instrument is in a live dialogic loop with the performer, who is looking for mastery of the instrument in the pursuit of the pleasure of playing and making the music happen. Instruments, by nature, are not immersive. There must be a triangulation between the player, the instrument and a visible or an audible result.

What is an Apparel?

A recent anthology published in French, *L'art au temps des appareils* (Huyghe 2005),[5] provides the occasion to introduce the notion of 'apparel' to further distinguish instruments from apparatus (*dispositif*) and apparel (*appareil*). In what follows, I will distinguish two versions of the apparel, one with a capital 'A' designating a system, and one with a small 'a' as a specific device. As pointed out earlier, the term *apparel* is not used much in current English, at least not in discourses on art. However, I would like to introduce it here as a potentially useful categorization of certain types of electronic and digital systems. In 1993, the Metropolitan Museum of Art had an exhibition entitled 'Infra-Apparel', an exhibition devoted to under-garments, which inspired me to use this term.

Interestingly, the English verb 'to apparel' is derived from the Middle English *appareillen*, from the Middle French *apareillier*, to prepare; it also derived from vernacular Latin *appariculare*, and from Latin *apparare*: to put clothes on. In this sense it is close to the noun 'implement', that I have discussed earlier, in that it designates an article serving to equip, 'the *implements* of religious worship'. The noun *apparel* means the equipment (as sails and rigging) of a ship; personal attire, something that clothes or adorns. In French, we find a rich network of terms derived from or part of the same family as the noun *appareil* and the verb *appareiller*: *apparat* (ceremony); *apparaux* (boat); *appareillage* (masonry, marine, boat, to go); *apparier* (two, to link). The semantic networks of these terms relate them to notions of *apparaître* (to appear), of *apparence* (appearance), which, as we know, can be both deceiving and revealing; they include ideas of joining of parts and of garment, of externalization and of departure.

According to Véronique Fabbri, the distinction between apparel and instrument lies in the study of the relations of apparel with materials (*matériaux*) structures and rhythms (Fabbri 2005: 96). In a passage where she conflates indistinctly many terms, she further tries to make distinctions:

> [T]he relation of the apparel to materials, thus distinguishing it from the instrument and the machine: the instrument, the tool, the machine have the common function of transforming a material, of submitting it to a form. The apparel, on the contrary, arranges the material and renders it available for transformation or for being set in motion (*mis en oeuvre*). (Fabbri 2005: 95)

Fabbri continues and points to Adorno's important affirmation of the concept of material in art, which has put in crisis the relation of the form and the content. Artistic works are less concerned with creating forms than with exploring those already in motion in the material.

The distinction we find here, badly expressed and confused, using the terms 'instrument', 'tool' and 'machine' without sufficiently defining them, seemingly grouped together in opposition to the apparel, must stop us for a moment. We could multiply examples of this kind of confusion. But let us retain here that the apparel seems to be what makes materials useable, conforms to a 'project', says Fabbri. The apparel sets up material potentialities and makes them available for human use, while the instrument transforms and informs the material. In electronic music, writes Fabbri, electronic audio systems and devices, which constitute the apparel of the studio, make possible for the sound to be a material at the composer's disposal, allowing him or her to distend, expand or contract sounds, to modify sound textures and orchestrate time in rhythms. Here, the Apparel of the studio and instruments in the studio share the material of sound (or electronic signals sent through audio channels); while the Apparel of the studio makes sounds available as material for composition, performance with electronic musical instruments forms and informs the material, here the sonic material. As long as the Apparel of the studio sustains the material of sound in its unformed and raw status, available and virtual, it is not involved with human intentionality.

The point here is not so much that instruments form and inform materials, rather that human intentionality is in the playful mode of instrumental playing. The use of instruments is guided by intentionality of the instrumental composition and the choice of instrumentation. Instruments also require instrumental gestures to be activated. Among instruments at the disposal of artists and composers, we can imagine a gradation from the pure apparel that turns on electrons to the very active gesture-triggered musical instrument, passing by the more intricate case of the instrument that has a certain number of automatic settings and combinations that requires minimal human intervention while performing or once set in motion.

I have used Apparel, with a capital 'A', to designate a system, like the sound studio, an addition or an interconnection of devices that may or may not include instruments. While instruments are not immersive, always maintaining or requiring the triangulation of the player, the instrument and the audible or visible results, the Apparel can be immersive, environmental, somehow abolishing the distinction of the user/player through participative/immersive modes. In this vein, one can use, as does Fabbri, the notion borrowed from Benjamin of the reception in distraction, the form of reception Benjamin sees as our relation to architecture, rather through habits and in a tactile and kinetic fashion than through distant contemplation and visual apprehension. Thus a distinctive mark of the instrument is its active and singular implementation of imagination and anticipation in performance within the Apparel.

The body appareled

The Apparel is immersive: it is an environment, it makes the signals lurk around us. It can even generate anxiety or anguish as 'big TV studios' did to Nam June Paik: 'Big TV studio always scares me. Many layers of "Machine Time" running parallel, engulf my identity' (Paik 1973). Another version of the apparel is also possible, this time with a small 'a', to distinguish it from the Apparel as system or ensemble. In this instance, the apparel is what adorns the body of the player/dancer/spectator: data suit, harness, head-mounted display, and the like; they become 'cyborgs' (Hables 1995). The body is here appareled; it is immersed in data feedback loops and, as the Apparel, this apparel is abolishing the distinction between the body and the 'world', and it favors tactile apprehension over distant visual or aural perception.

Such distinctions between the instrument and the apparel help us to have a more nuanced comprehension of how artists and composers invent and use technological devices. These distinctions are also necessary to respond to the numerous art projects with virtual reality, augmented reality or responsive environment. With these new, often immersive environments, the notion of a body-appareled has to be distinct from the body-playing-instruments, from instrumental playing, even though they all use similar technologies. A body-appareled does not necessarily (perhaps with the exception of that of a dancer) control the apparel, does not necessarily search for mastering it, and might as well be more inclined toward surrendering to the machine; instrumental playing relates the body to an object, an instrument responding to gestures that produce a certain set of sonic or visual output. The effect of one is immersion and is experiential, tactile; the effect of the other is pleasure at mastering the production of sonic and visual forms as aesthetic objects and as playing in performance.

What is an Apparatus?

Italian philosopher Giorgio Agamben recently published in French a book entitled: *Qu'est-ce qu'un dispositive?* (2007) ('What is an Apparatus?') in which he establishes the etymological source of the term *apparatus*. Agamben questions this concept used by Foucault, who never really defined it. The philosopher retraces briefly the origin of the term, as employed by Foucault, from his philosophy professor Jean Hypolite under the concept of 'positivity'. Hyppolite interprets the Hegelian concept of positivity as the historical anchorage of religion and as constraints to human liberty, as imposition of institutional norms onto the individual beliefs and subjectivity. Indeed, all apparatuses have to do with the construction of the subject and his/her position relating to a concrete and particular situation: '*Ils doivent produire leur sujet*' ('They must produce their subject') (Agamben 2007: 27).

In French, the term *dispositif* is often encountered in contemporary art discourse. In English, *apparatus* is probably less in use outside cinema criticism; one would see 'device' as the more common term which shares with the French *dispositif* a vagueness as to what it is precisely. Devices are often part of and confused with installations, as the device generally designates any assortment of electronic or digital equipments, intervening in

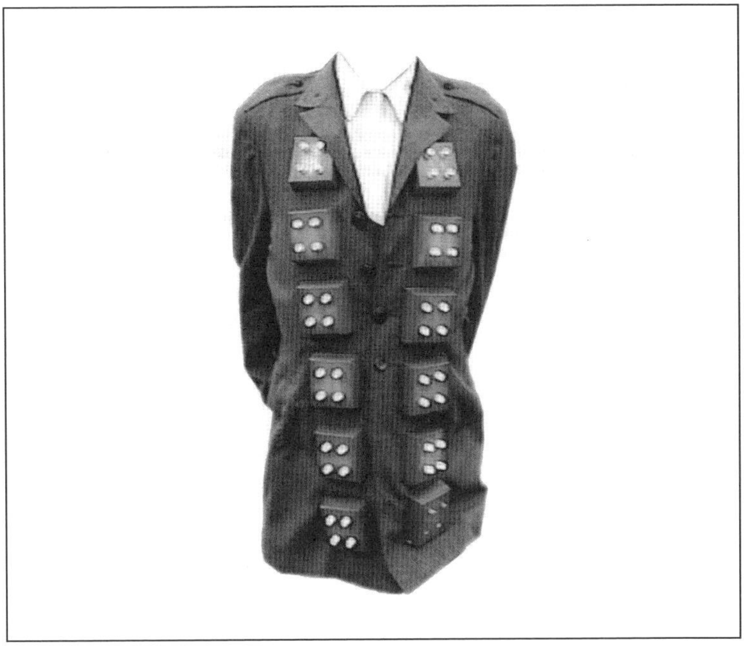

Figure 2. *Le manteaux de l'admiral (the admiral's jacket)*, Peter Blasser, wearable instrument (2003). Image from Peter Blasser documentary file, Daniel Langlois Foundation de Cinémathéque québécoise. (courtesy Cinémathéque québécoise)

the space or in the relation of the spectator with the image and with his or her self-image, to transform the experiencing subject and the space of the work.

The cinematographic apparatus

Not known for much else in film studies but for two articles published during the first half of the 1970s – a decade that saw the bringing together or the merging of Lacanian psychoanalysis, literary semiology and Marxist philosophical inquiry into ideological analysis, first in France, and then in the Anglo-American universities – Jean-Louis Baudry was at the inception, with Christian Metz and others, of a particular way of analyzing cinema. The aim of Baudry when he wrote 'Ideological Effects of the Basic Cinematographic Apparatus' (1970) and 'The Apparatus: Metapsychological Approaches to the Impression of reality in Cinema' (1975)[6] was to decode the technical film apparatus in terms of an ideological configuration meant for replacing the comprehension of reality (human, material, cultural and economic) by the virtue of misrecognition, suspension of disbelief and the impression of reality. It is also one limit of Baudry's texts in that they don't look outside the mainstream situation of the fiction feature film; the author in fact does not talk about any particular film in order to focus on the canonic commercial cinema experience. His object is the reception, the spectatorship positioning in relation to the basic technology

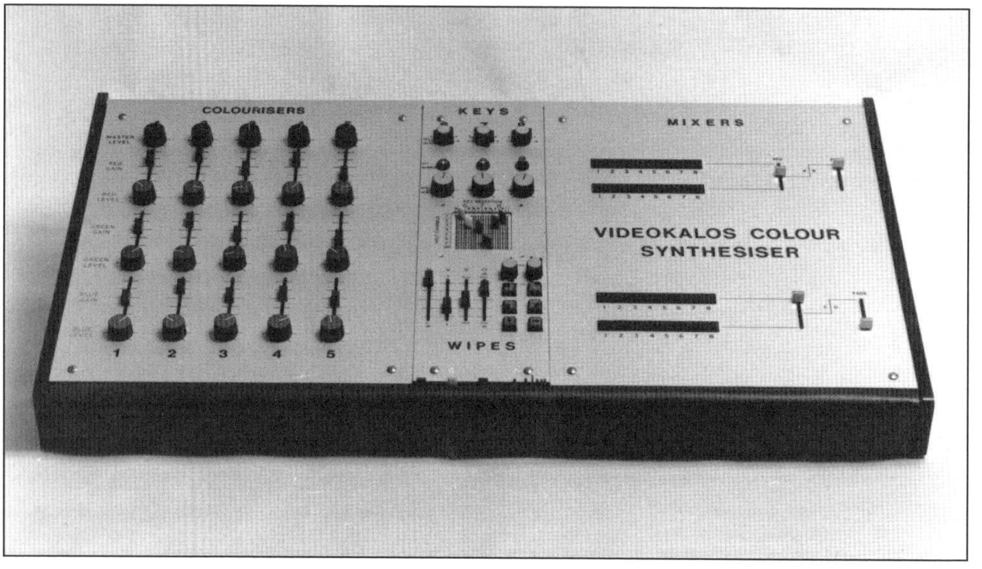

Figure 3. Videokalos color synthesizer; engineer. Peter Donnebauer (c. 1978). Image from Steina and Woody Vasulka Fonds, VAS B38-C11-3, Collection Fondation Daniel Langlois de la Cinémathèque québécoise, (courtesy. Cinémathèque québécoise).

of cinema: the projector at one end of a darkened room with the shadows on the screen at the other end, with the spectators in the middle 'harnessed' to their seats, *sujets tout percevant*, all-perceiving subjects, as Christian Metz would have said.

The impact that the notion of apparatus had on Anglo-American film studies has been mainly through the translation of Althusser's text. The term apparatus had been introduced in Marxist philosophy by Louis Althusser as in the 'Ideological State Apparatus'. It had a good English *fortune critique* through two books: the first, *Apparatus, cinematographic apparatus: Selected Writings* (Cha 1980) and the second, *Narrative, apparatus, ideology: A Film Theory Reader* (Rosen 1986). This is not exactly a novelty, and anyone familiar with 1970s French cinema studies is familiar with this kind of ideological analysis. But Baudry's concept of the apparatus was expanded later in the 1980s by Anne-Marie Duguet. Baudry concentrates on the question of ideological analysis with an Althusserian undertone; in his second text, he looks into the psychoanalytical basis of the impression of reality, this time in Lacanian terms. Of course, in the end, the two texts are somewhat linked, since ideological delusion functions in similar ways as the impression of reality: both function through psychic projections and identifications, through misrecognitions and misconceptions. Baudry's notion of ideology here is not concerned with discourses but with their vehicle. His idea of ideology is not a matter of promoting a political discourse or denouncing one; it is rather to analyze how the spectator's positioning installs him or her[7] in a receptive posture or mode allowing these discourses to be effective, to have an effect in return on the spectators' lives and structures as subjects. Baudry defines the apparatus of cinema as 'support and instrument of ideology [which constitutes] the "subject" by the illusory delimitation of a central position' (Rosen 1986: 295), in view of providing for the maintenance of a certain social order and of idealism.

The notion of apparatus became more largely used outside of film studies to consider artworks that used some sort of technological devices (video, audio) for presenting moving images and sounds. Anne-Marie Duguet was one important proponent of the *dispositif*, of the concept of apparatus in a larger sense. But this was done to the detriment of ideological analysis, which disappears behind the curtain of 'critical' artistic practices and forms within the art discourse. In her article entitled 'Dispositifs' (Duguet 2002: 42), Duguet notes that video installations, using an electronic apparatus (*dispositif électronique*), activate a radical displacement of the experience of the work, and that the work is a relational system which returns the spectator to his own perceptive activity (Duguet 2002: 17). Recognizing her debt to Baudry's work as well as those of Christian Metz and Thierry Kuntzel, she points out the definite fact that video and electronic (nowadays, digital) systems allow artists greater liberty in the arrangement of elements in the work, playing on the malleability in capturing, producing, reproducing, disseminating and perceiving images and sounds, by imagining such apparatuses for reflecting on spectatorship within the work. This opens up to interactivity, but the apparatus is still a device or set of devices aiming at decentering or displacing the spectator, and dislodging him or her from the position of stillness that cinema seemed to impose.

By pointing to differences between the Apparel and the Apparatus, this later concept, as I just too briefly exposed, appears to have historical contours by which its terms of reference, Althusserian ideological analysis and Lacanian psychoanalysis, are concerned with a visual field rather than a tactile or an aural one and, consequently, it conceives spectatorship in terms of the centralized/decentralized perceiving subject. I have shown that the notion of instrument and of instrumental playing brings about an active and gestural embodiment that demands a degree of knowledge and mastery as opposed to a passive relation to an apparel or being surrounded by the Apparel. Both instruments and apparels might be part of the Apparatus of a work: both are part of the positioning of the subject within a work, but they certainly differ in the way they establish this position. The construction of the subject differs greatly whether you play an instrument with a distancing triangulation at the heart of the instrumental relation or you are immersed in or surrounded by the Apparel where boundaries of subject and object tend to blur.

References

Agamben, Giorgio (2007), *Qu'est-ce qu'un dispositive?*, Paris: Payot et Rivages.

Baudry, Jean-Louis (1970), 'Ideological Effects of the Basic Cinematographic Apparatus', *Cinéthique*, 7/8.

—— (1975), 'The Apparatus: Metapsychological Approaches to the Impression of reality in Cinema', *Communications*, 23.

Cha, Theresa Hak Kyung (ed.) (1980), *Apparatus, cinematographic apparatus: Selected Writings*, New York: Tanam Press.
Curtis, James M. (1978), *Culture as Polyphony: An Essay on the Nature of Paradigms*, Columbia, MI & London: The University of Missouri Press.
Dufourt, Hugues (1991), *Musique, pouvoir, écriture*, Paris: Christian Bourgois Édietur.
Duguet, Anne-Marie (2002), *Déjouer l'image, créations électroniques et numériques*, Nîmes: Jacqueline Chambon. First published 1988.
Fabbri, Véronique (2005), 'De la structure au rythme. L'appareillage des corps dans la danse', in P. Huyghe (ed.), *L'art au temps des appareils*, Paris: L'Harmattan-Esthétiques, pp. 93–121.
Gagnon, Jean (2005), 'Media phantasmagoria', in M. Tromble (ed.), *The Art and Films of Lynn Hershman Leeson: Secret Agents, Private I*, Berkeley: University of California Press; Seattle: Henry Art Gallery, pp. 139–45.
Hables, Chris Gray (ed.) (1995), *The Cyborg Handbook*, New York & London: Routledge.
Huyghe, Pierre-Damien (ed.) (2005), *L'art au temps des appareils*, Paris: L'Harmattan-Esthétiques.
Ihde, Don (1991), *Instrumental Realism: The Interface between Philosophy of Science and Philosophy of Technology*, Bloomington, IN & Indianapolis: Indiana University Press.
——— (2007), 'Technologies – Musics – Embodiments', *Janus Head*, 10: 1, pp. 7–24.
Leroi-Gourhan, André (1971), *L'homme et la matière*, Paris: Albin Michel. First published 1943.
——— (1973), *Milieu et technique*, Paris: Albin Michel. First published 1945.
Miranda, Eduardo Reck and Wanderley, Marcelo M. (2006), *New Digital Musical Instruments: Control and Interaction Beyond the Keyboard*, Middleton, WI: A-R Editions.
Paik, Nam June (1973), *Videa 'n' videology 1959–1973*, Syracuse, NY: Everson Museum of Art.
Rosen, Philip (ed.) (1986), *Narrative, apparatus, ideology: A Film Theory Reader*, New York: Columbia University Press.
Schaeffner, André (1994), *Origine des instruments de musique, Introduction ethnologiques à l'histoire de la musique instrumentale*, Paris: Éditions de l'École des hautes études en sciences sociales.
Simondon, Gilbert (1989), *Du Mode d'existence des objets techniques*, Paris: Aubier. First published 1958.

Notes

1. The famous studies of André Leroi-Gourhan on the relationship of humans to tools and to their surrounding material milieu are eloquent on this point. See Leroi-Gourhan (1971; 1973).
2. Even though Simondon's book was first published in French more than 50 years ago, there is still no full English translation of it. You can find online part one of the book in English translation: http://accursedshare.blogspot.com/2007/11/gilbert-simondon-on-mode-of-existence.html. There is also useful information about him on Wikipedia : http://en.wikipedia.org/wiki/Gilbert_Simondon. All quotes from this book are my translation.
3. Playing and games, like the arts, are activities and things without finality other than their own performance or representation, generating symbolic pleasures and satisfactions for those who are participating in or witnessing them. At least that would be Hans Georg Gadamer's argument concerning the auto-representation of games.
4. See Curtis (1978).
5. The title could read in English: *Art in the Time of Apparels*. All quotes from this book are my translation.
6. A third text, which is not discussed here is entitled 'Author and Analysable Subject', and appears in Theresa Hak Kyung Cha (1980).
7. Feminist film analysis has uncovered this spectator's positioning as not being the same whether a man or a woman is seated in the theater. See my text, 'Media phantasmagoria' (Gagnon 2005).

Expanding 'Image-processed Video' as Art: Subverting and Building Control Systems

Jeremy Culler

> Alternative television also meant creating images that looked different from the standard T.V. Thus, 'image processing' as we know it grew out of an intensive period of experimentation that for some, in a vague way, was seen visually to subvert the system that brought the Vietnam War home every night. (Furlong 1983a: 35)

In 'Electronic Video Image Processing: Notes Towards a Definition', Sherry Miller Hocking, Assistant Director of the Experimental Television Center, describes the process of electronic image processing as using 'properties inherent in the medium of video' as 'art-making material' (Miller 1983: 22). Hocking explains:

> Artists work at a fundamental level with various parameters of the electronic signal, for example, frequency, amplitude, or phase, which actually define the resulting image and sound. Electronic tools are the instruments with which the signal is created and then altered. These signals carry the image in an electronic coded form. These coded structures are what the artist actually works with when creating an image-processed work. When these signals are decoded by a television monitor, the images and sound are displayed. (Miller 1983: 22)

Hocking's statement about the mechanics of electronic image processing not only explains a technical process that uses inherent properties of the video medium as 'art-making material', but also describes what artists do to time-based and time-dependent signals in order to create artworks for television (Miller 1983: 22–23). As a technique, image processing involves manipulating the various parameters of the electronic signal, including frequency, phase and amplitude, which, as Hocking notes, 'actually define the resulting image and sound' (Miller 1983: 22). Artists working with these parameters are able to change the electronic configuration of the standard television signal by feeding it through raster manipulation devices, colorizers, mixers and/or audio and video synthesizers. They also advance what Hocking and others see as a distinct genre, known as 'image-processed video', within video-based art.[1]

In principle, then, 'image-processed video' seems like an appropriate term for describing this type of work. However, many artists and scholars argue that the term is misleading. This is because, as Lucinda Furlong notes, it may be used to reference all works on television containing synthesized imagery, including music videos created for commercial television networks, such as MTV, and time-based correction procedures employed by broadcast television engineers (Furlong 1983a: 35). For this reason, Furlong considers 'image-processed video' a pejorative and misleading term (Furlong 1985: 233). She explains that in addition to its use as 'a genre and a catch-all phrase for every technical process in the book' the term 'conjures up a number of very specific – often pejorative – stereotypes: densely layered "psychedelic" images composed of soft, undulating forms in which highly saturated colors give a painterly effect, or geometric abstractions that undergo a series of visual permutations' (Furlong 1985: 233). For many artists using image-processing tools and techniques, these characterizations, which emphasize non-aesthetic results that television engineers learn to avoid in commercial broadcasting, are 'superficial and belie the range of concerns that fall within the image-processing umbrella' (Furlong 1985: 233). Despite such stereotypes, however, artists continued to develop new electronic devices and image-processing techniques and, as a consequence, advanced the idea of 'image-processed video' as art.

The idea of 'video image-processed art', according to Furlong, germinated out of 'an intensive period of experimentation that for some, in a vague way, was seen *visually* to subvert the system that brought the Vietnam War home every night' (Furlong 1985: 234). Artists – including Stephen Beck, Bill and Louise Etra, Ralph Hocking, Nam June Paik, Steve Rutt, Dan Sandin, Eric Siegel, and Steina and Woody Vasulka – used image processing not only to subvert commercial television programming, but also to explore 'the electronic signal as a plastic medium, a material with inherent properties that can be isolated' (Furlong 1985: 234).[2] For instance, Steina and Woody Vasulka utilized custom-built equipment to isolate and then construct video frames as plastic, compositional material in works such as *Matrix I* (1970–72) (Figure 1). This formalist approach was one example, among countless others, that turned out to be 'central to the development of what became the image-processing aesthetic' for artists (Furlong 1985: 234).

It is clear then why many artists see 'video image processing' as an inadequate term. That the term could compromise the integrity of artworks and, thus, problematize the validation and legitimization of electronic video image processing as an innovative art genre is why so many reject it. The term's problematic signification, however, did not hinder the advancement of the idea of 'image-processed video' art at alternative resource centers, public (access) television stations, galleries and museums during the 1970s and 1980s – nor did it stop scholars and artists from using the term in academic texts or alternative production manuals (Furlong 1983a: 35–38; 1983b: 12–17; 1985: 233–37; and Tamblyn 1991: 303–10). A determination to expand television and video as art forms led artists and engineers, who wanted to challenge commercial television programming, to develop an 'image-processing aesthetic' and establish a visual language unique to video-based art.

One of the primary ways that artists advanced this new visual language was by using pioneering 'control systems' and image-processing devices developed at the Experimental Television Center (ETC). These 'control systems', which directors Ralph and Sherry Miller Hocking define in an ETC funding report (1980–81), are 'either manual or pre-programmable in operation' (Hocking and Hocking 1981). They explain:

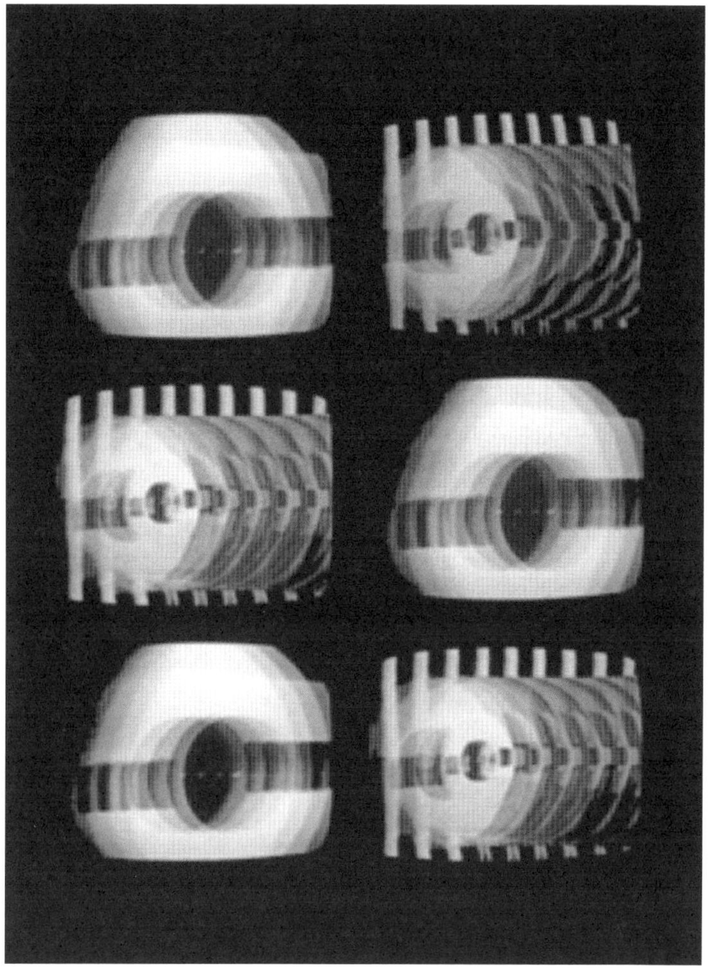

Figure 1. Steina and Woody Vasulka, detail of *Matrix I* (1970–72).

The adjustment of a knob setting by hand to change hue, for example, is the simplest example of *manual control*. This is primarily a gestural and reactive process and although the apparent result is an alteration of the compositional element color, the artist is actually manipulating specific parameters of the electronic video signal which defines the change in the image. The analog control device generates waveforms which can be used as image and sound and as control signals. With this method of control, the manual adjustment is replaced by a preconceived and structured signal voltage which causes the change in the compositional element. The computer control system [on the other hand] provides to the artist methods of *pre-programming* the image changes; the

> result is the codification and, therefore, the precise repeatability of the process. (Hocking and Hocking 1981; my emphasis)

Whereas manual control systems involve a process that is gestural and reactive – requiring a time-dependent, manual adjustment of a knob for example – a computer control system offers preprogrammed options that process a signal internally. For example, in using the image grabber or frame buffer function in a Z-2 8-bit microprocessor, one can capture a video image and convert it into discrete blocks of 16 shades of gray (Hocking and Hocking 1981). Hocking explains that in addition to being able to reorganize or group the gray levels in the original image, 'this computer system can control pre-defined image changes, for example color, and can also translate camera or other input images into digital code, buffering the image and operating on the signal code to define the image in terms of gray levels' (Hocking and Hocking 1981).

Integrating analog and digital video image-processing devices into the Experimental Television Center's studio and building manual and preprogrammable control systems to operate them was only part of what was needed to enable artists, educators and members of the community to produce artworks and advance 'image-processed video' as art. The other component involved pedagogy. In a report discussing workshops available at the Experimental Television Center between 1972 and 1981, its directors emphasized their commitment to educating the public by offering courses, such as 'Basic Video', 'Video Post-Production', and 'Image Processing and Video Art', for free (Hocking n.d.). These courses provided instruction on topics that were unavailable to the public at other institutions, such as public television studios (Hocking n.d.). They also expanded Ralph Hocking's idea for developing a space that offered television and video production resources to artists. This idea began in 1969, when Hocking founded Student Experiment in Television (SET) at the State University of New York, Binghamton and continued in 1970 and 1971, when he established the Community Center for Television Production (CCTVP) and the Experimental Television Center, respectively.

After 1971, the Experimental Television Center expanded its mission and, in turn, cultivated social networks, established community research and screening programs, and developed an analog/digital hybrid studio. Many in the field also shared the Center's additional interest in subverting and building control systems, not for profit or notoriety, but instead for the chance to facilitate the advancement and expansion of image-processed video as a viable mode of cinematic and artistic expression. On the one hand, the advancement of electronic video-imaging systems was crucial if image-processed video was to be considered more than just a technical craft. On the other hand, cultivating networks, building control systems and facilitating technical processes that use the inherent properties of the video medium as art-making material were essential if the development of a visual language unique to the video-based art genre was to occur. That the discursively loaded term 'image-processing' can be used as 'a catch-all phrase for every technical process in the book' (Furlong 1985: 233) does not mean that it nullifies the real value of approximately four decades of media arts using image-processed video. Rather, it suggests that from the very start, image-processed video was a heterogeneous phenomenon that, for pioneering artists, provided innovative solutions for subverting the restrictive nature of commercial television. In other words, image-processed video was a means with which artists could exceed the limits of commercial television programming and expand the boundaries of television and video production for artistic and aesthetic purposes.

References

Furlong, Lucinda (1983a), 'Notes toward a History of Image-processed Video: Eric Siegel, Stephen Beck, Dan Sandin, Steve Rutt, Bill and Louise Etra', *Afterimage*, 11: 1–2.

—— (1983b), 'Notes toward a History of Image-processed Video: Eric Siegel, Stephen Beck, Dan Sandin, Steve Rutt, Bill and Louise Etra', *Afterimage*, 11: 5.

—— (1985), 'Tracking Video Art: "Image Processing" as a Genre', *Art Journal*, 45: 3, pp. 233–37.

Hocking, Ralph and Hocking, Sherry Miller (1981), *Funding Report 1980–81*, Newark Valley, NY: Experimental Television Center Archives.

Hocking, Sherry Miller (n.d.), 'History of the Experimental Television Center and SUNY: Workshops', Newark Valley, NY: Experimental Television Center Archives.

Miller, Sherry (1983), 'Electronic Video Image Processing: Notes Towards a Definition', *Exposure*, 21: 1.

Tamblyn, Christine (1991), 'Image Processing in Chicago Video Art, 1970–1980', *Leonardo*, 24: 3, pp. 303–10.

Notes

1. For example, see Furlong (1985).
2. It should be noted that the 'material' that Furlong refers to is actually de-material or immaterial in nature.

The Grammar of Electronic Image Processing

Sherry Miller Hocking

Video is an art of space/time. As artists first adopted video technology to create art in the late 1960s, they struggled to understand how the technology generated and altered images, and the implications of those processes for their work. It was clear that video was fundamentally different from film. The motion of film was created by rapidly viewing a series of photographed stills; holding the film up to the light revealed each of those frames as discrete images. The video image did not exist as an entity at all. Inspecting recorded videotape reveals nothing. The tape holds signals which have encoded images as electrical impulses playing out point by point through time. Film is projected and we experience the motion as reflected light. Video is displayed in such a manner that we look directly into the light source. While both media present motion, it was clear from the beginning that video was unique in its construction. What were the implications of the processes of video to the aesthetics of video art practice?

For many early makers, an understanding of the basic language of the technology of image creation – the grammar of the electronic image – was necessary:

> Film was boring while it imitated the conventions of the proscenium arch stage; and television remains trivial while it imitates film. Artists and innovators signal their break with such trivial use of television by calling *their* use 'video'. Video rejects the conventions of both film and broadcast television and attempts to discover the unique formal necessities of its electronic processes. Video is finding the conventions suitable to such necessities; and we now have an electronic visual art form to complement electronic music. (Arn 1973: 15; original emphasis)

As artists took up the tools of video making, many began to construct a language which could not only explain the technical processes, but offered tools for creating a new aesthetic understanding of this new medium.

Woody Vasulka referred to the 'organizational models' of the electronic image in 1975:

> The aesthetic here is specifically didactic: to visually display, as clearly as possible, the step-by-step development of very primitive, basic modes of information available with this [the Rutt/Etra] synthesizer. Accordingly, sine, triangle, or square waves are used as the bases for most images. The didactic purpose involved is to enable the principles of time-energy construction to become common knowledge, as a primary

conceptual and technological tool of our evolving electronic society. (Vasulka and Nygren 1975: 9)

Vasulka goes on to describe his study of smaller and smaller time-sequences of the video signal, necessary to an understanding of the formation and programmability of waveforms. In Barbara Buckner's *Light and Darkness in the Electronic Landscape: Some Aspects of the Video Image* presented at the Collective for Living Cinema in 1978, she also remarked about this grammar of video.

In the early 1970s, people at ETC began to write a series of manuals which attempted to define the basic vocabulary of video – the formal characteristics of the medium. We tried to articulate the building blocks of the image, and to categorize each of the characteristics and processes in terms of functionality within a larger system. Peer Bode, Richard Brewster, Hank Rudolph, Matthew Schlanger and Walter Wright among others contributed to the writings of the ETC texts, and have engaged in their own studies, both text- and image-based, about the foundational principles of the video image.

The image-processing system was used in a wide variety of applications, from real-time documentary recordings, to imagistic narratives, to formal abstractions. They were presented as live performances, prerecorded tapes, interactive sculptures and installations. Regardless of the form the art took or the specifics of the system employed, the image-processing systems could be seen as grounded in a similar grammar: the grammar of light in motion.

The definitions which follow are taken primarily from manuals written between 1974 and 1985 at ETC.[1] Portions were reprinted in *Eigenwelt der Apparatewelt: Pioneers of Electronic Art*. The text focuses primarily on analog signals. A discussion of digital and computer tools is included elsewhere in this section.

Signal

The generation and display of the video image are time dependent. The composition of the signal defines the visual nature of the image as it exists in time; it dictates both the appearance of the single 'still' image, which really exists within a specific duration of time, and its behavior through time.

On a basic level, the signal can be viewed as the art-making 'material'; the creation of an electronic image is an architectural process and constructed in time. Video images are codes of information conveyed by signals. The specific video picture information conveyed by a signal is in the form of changes in voltage; changes in voltage dictate changes in the information being carried, and thus the image and sound. In this way the hardware of the system can be viewed, in part, as 'a carrier of aesthetic definitions' (Hagen and Kite 1978: 22).

Signals within a processing system can be categorized as video, audio, control, and synchronizing or timing signals. Signals may function in single or multiple categories. For example, a signal can influence aspects of an image – a process called voltage control – and also produce a sound.

Two devices are used to check the output signal coming from a processing system: a waveform monitor and a vectorscope. A waveform monitor does no processing of the video signal, but it allows us to examine the video signal by displaying a graphic representation of the changes in the voltage of the video signal over time. The waveform monitor is really a special-purpose oscilloscope. Vertical distance on the waveform display represents voltage, while horizontal represents time. One field or one line of the video signal can be observed. Each aspect of a video signal must fall within a predetermined set of limits so that the recorded tape is compatible with other playback machines and can be displayed. The waveform monitor ensures that the system is operating within these prescribed parameters.

A vectorscope shows the color portion of the video signal and ensures that the recorded colors are consistent. It uses the convention of a color wheel to represent the signal. Chroma, or saturation, is indicated by how far the signal extends from the center. The hues are marked at specific points initialed M (magenta), R (red), G (green), Y (yellow), B (blue) and C (cyan). At a specific setting, when color bars are patched to the vectorscope input, the signal's six points will correspond to the correct marks.

Signals can be further specified as analog or digital structures; the terms refer to how changes occur, either discretely or continuously. An analog signal is a voltage which continuously varies within its allowable range. A digital signal consists of discrete levels or parts. Digital signals are concerned with stepped information; the change from one value to another in a waveform is not continuous but, with some qualification, is instantaneous and the signal is either on or off.

A signal then conveys certain information about an event. It contains a number of variables, such as frequency, amplitude or placement, which can be changed and controlled.

The inputs to an image-processing system are signals generated by video cameras, sonic modules, the output of another processor – for example, a keyer – or in certain cases prerecorded video.

Scanning

In film the impression of movement is derived from a succession of frozen moments. In contrast, the video image, even if each frame is examined, is all motion – a single rapidly moving and constantly changing dot, one dot only, does all the work. The basic illusion of film is motion. The basic illusion of video is stillness. A detail of the video image may be located by pointing out where it is (as in film), but also by specifying its distance in time from any other point of the image. Any point on the image is both 'where' and 'when' or 'wherewhen' from any other point. Video is quite literally a space/time machine. (Arn 1973: 17)

If the motion we attribute to the film image is as illusion, nevertheless the serial still frames of cinema are discretely apprehensible entities that may be held in the hand and examined at our leisure. When these frames are projected, they are uniformly interwoven with equal intervals of total darkness, which affords us intermittent moments to think about what we have just seen. Conversely, the video field is continuous, incessantly growing and decaying before our eyes. Strictly speaking, there is no instant of time during which the video image may properly be said to 'exist'. Rather, a little like Bishop Berkeley's imaginary tree – falling forever in a real forest – each video frame represents a brief summation within the eye of the beholder. (Frampton 1977: 34)

The raster is the visible rectangle of light coming from the cathode ray tube (CRT) of a monitor or camera when no picture information is displayed. Scanning describes the process by which the raster is constructed in the CRT of both camera and monitor. The function of the scanning process in a camera is to 'read' the object before it as a mosaic of varying light and dark values, converting them to variations in the electrical signal so they may be reproduced on a monitor in real time, further processed by an electronic processing system or recorded by a videotape recorder for storage and later display. The function of the scanning process in a monitor is to display the image of the object presented to the camera or derived from the output of a processing system. The camera thus translates the scene before it into a signal which reproduces these fluctuations in light and dark; light energy is thus changed to electrical energy. The function of the monitor is to translate the electrical signal containing the image information into a perceivable image; in the monitor, electricity is converted to light energy.

In both the camera and monitor, the raster is constructed by a beam of electrons which is focused to a point; this point is moved in a continuous and repeatable manner so that the rectangular field of the raster is inscribed. The scanning process is linear and sequential; the electron beam excites a single unit at a time across each line, called a 'picture element' or 'pixel', which emits light when hit. The beam then progresses to the next pixel. In this way a horizontal line is written; as they stack up the raster is formed. The raster is a matrix of dots. Each field is made up of 262 1/2 lines, or exactly half the number of lines comprising one frame. The first field is composed of all the odd numbered lines, the second field the even. Each frame is composed of two separate fields or 525 lines. If the lines of the even field fall precisely between those of the odd field, this is known as 'interlace scanning'. One field is scanned every 1/60 second; since the frame is constructed of two separate fields, each scanned 60 times per second, one frame is displayed in 1/30 second or 30 frames per second. Because of the phenomenon of persistence of vision and the retentive characteristics of the phosphor surface of the cathode ray tube, we perceive complete images rather than the travels of a single dot. The serial nature of image formation is characteristic of video and very unlike film. In video, the entire picture exists only as a function of time and is never present as a single entity. The term 'frame' is somewhat misleading because of the association of a complete image which exists at any one instant of time. The video frame is a process of constructing and deconstructing the image.

Sync

The video signal has two basic parts: the section containing picture information; and the section containing sync information. The sync information is composed of a number of individual types of signals. Synchronization is derived from the Greek *syn* and *chronos* (to be together in time); the term implies that several processes are made to occur together in time at the same rate so that they are concurrent. For a coherent picture to be formed which is readable to the eye and brain, the scanning motions of both the image (signal) generating dcvicc – for example a camera – and the image (signal) display device – the monitor – must proceed in an orderly and repeatable manner. The scanning processes in both camera and monitor must begin and end at precisely the same time. The camera and monitor must be in sync.

As the camera begins scanning the objects in front of it, the monitor begins to scan the line which the camera is scanning. As the camera ends the scan line, the monitor must also end that line. When the camera reaches the bottom of the field, the monitor must be exactly in step. Without this synchronization, the camera image and the monitor image will have no relationship to each other. 'Horizontal sync' maintains the horizontal lines in step; without horizontal sync the picture will break up into diagonal lines. Horizontal sync dictates when

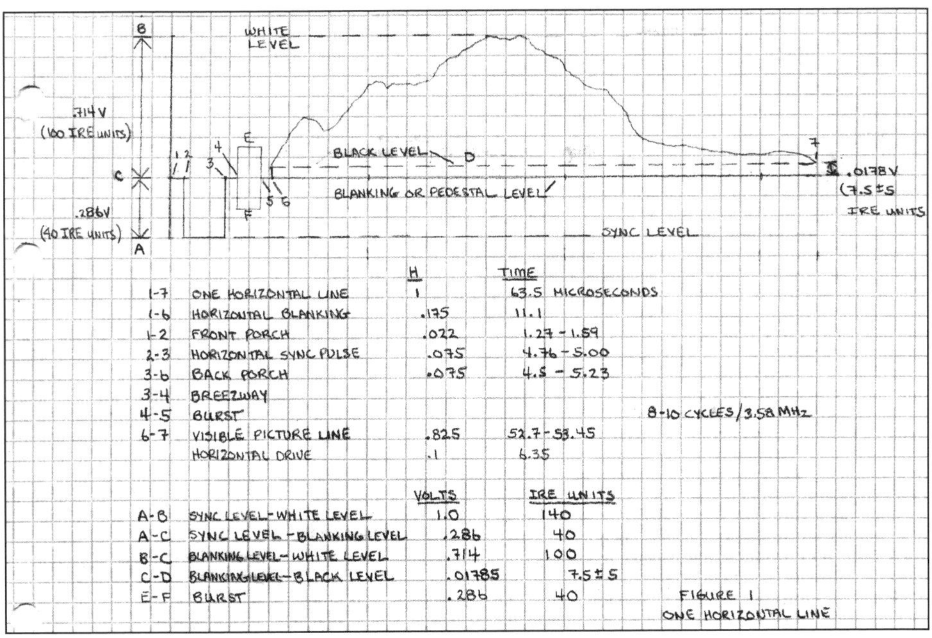

Figure 1. The video signal for one horizontal line, with sync by Sherry Miller Hocking (1986).

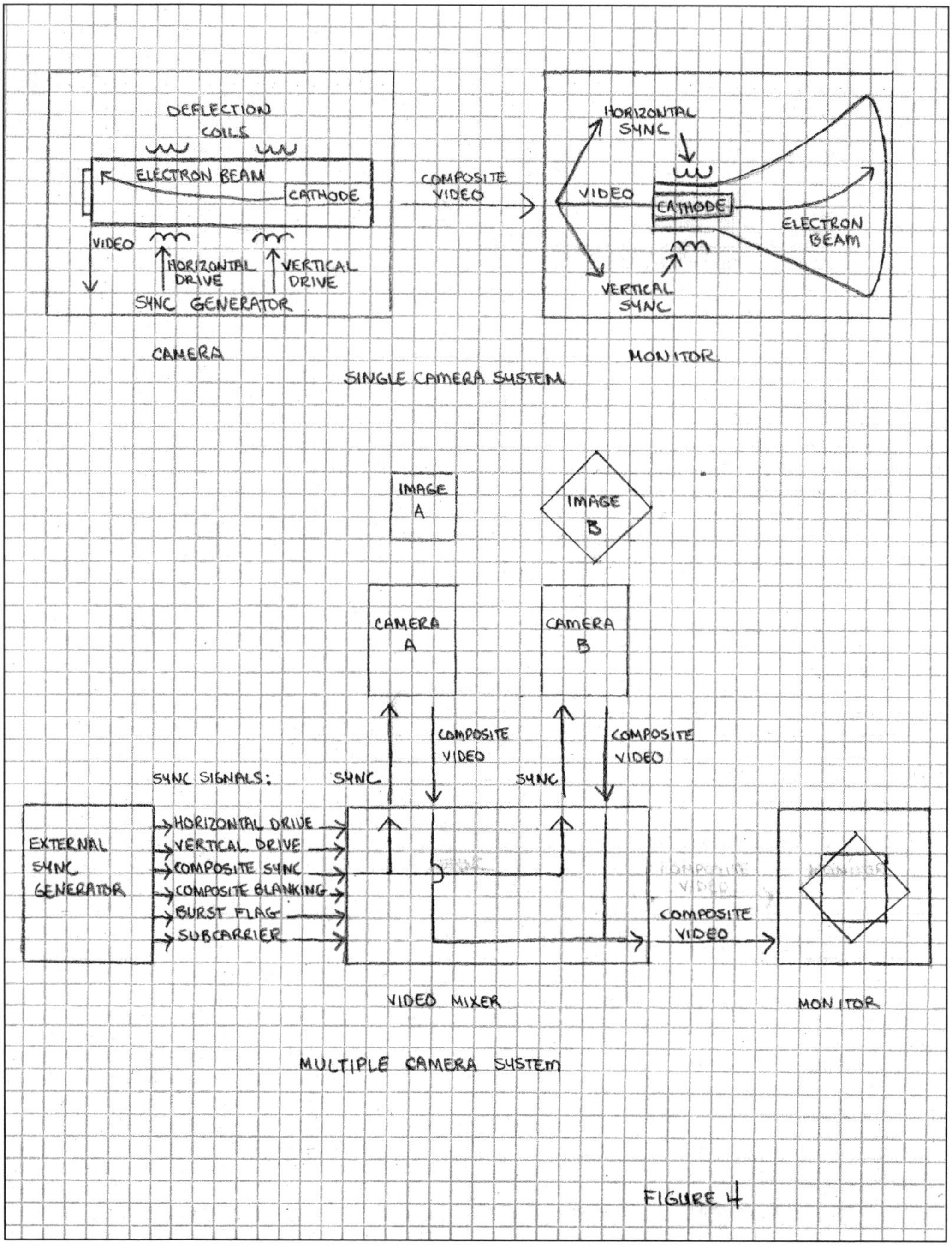

Figure 2. Sync and signal flow in single and multiple camera systems by Sherry Miller Hocking (1986).

each horizontal line begins and ends. 'Vertical sync' also keeps the picture stable; without this, the image will roll. Vertical sync tells the camera and monitor when each field begins and ends. Both together are essential to a stable rectangular shape. Sync then can be conceived of as an electronic grid which provides horizontal and vertical orientation to the image.

Each visible line forming the raster is drawn from left to right across the CRT. Before beginning to scan the next line, the beam must return to the left. During this horizontal retrace period, the beam is invisible; this process and the time interval necessary to perform this function are called 'horizontal blanking'. Horizontal blanking is a part of synchronization.

At the end of each field the beam must return from the bottom to the top of the CRT before beginning to scan the next field. Again, this vertical retrace is not seen. This process and the interval are referred to as 'vertical blanking'. Vertical blanking is also a part of synchronization.

Sync and drive pulses are the timing pulses which keep one or several cameras in step with each other and with the videotape recorder or monitor. In a single-camera system, sync can be obtained from the internal sync generator built into the camera. In a multiple-camera system, all cameras must receive the same sync signals from a common source at the same time, from a sync generator external to all of the cameras. Video or picture signals from all the cameras are then mixed in the processing system and combined with sync information. This single composite signal, containing both picture and sync information, is sent to the deck to be recorded. The sync generator serves as a master clock which establishes the time frames for the signals which, when decoded, produce images.

A color sync generator supplies horizontal and vertical drive, composite sync and composite blanking, and two additional signals variously called 'burst' or 'burst flag', and 'subcarrier' or 3.58 MHz. Color signals must carry all color information, including the hue, brightness and saturation of the colors, by the use of three primary colors: red, green and blue. In addition, their structure must be such that they are compatible with black-and-white systems. Color signals must therefore contain both luminance and chrominance information. Luminance conveys the variations of light intensity and is the part of the signal used by the black-and-white monitor. Chrominance conveys variations of hue, saturation and brightness. The subcarrier signal, a frequency of 3.58 MHz, carries information about color value. This frequency is produced by an oscillator in the sync generator, and is modulated or changed by the color information coming from the color camera to the colorizer in the image-processing system.

Sync also plays a role in a system which uses a prerecorded videotape as an input. In this instance, it is necessary to lock the system sync to the sync from the source videotape recorder. A genlock is required for this operation. Genlock is also used for cameras which are not externally syncable. This includes most consumer cameras. The output of the camera goes to the genlock input of the SEG and the system will lock to the internal sync of that camera. A VTR cannot be used as a direct source if the genlock is occupied by the camera.

Sources, processors and controllers

Video synthesizer and *image processor* are general terms referring to an assemblage of individual video signal sources and processors, all of which are integrated into a single system or meta-tool. The ETC system is designed to let the artist actually create an individualized system by patching various modules together to route the signals.

There are three general categories of devices: (1) signal sources – devices which output a signal used in the system to generate an image, a control signal or a sync signal; (2) processors – devices which perform some operation upon the signals, such as gain or phase changes, and are often used to mix inputs and put out processed signals; and (3) controllers – devices which generate signals which are themselves inputs to processing devices to control an aspect of the image. These devices and the signals can be analog or digital in nature.

Image generators produce optically based signals from cameras or non-optically based signals from oscillators. Image processors are devices which alter a signal from these sources in different ways; each processing module provides a means by which a single parameter or set of parameters of an image can be changed. Processing modules within a system are often arranged so that the output of one – a keyer, for example – can serve as an input for a second – for example, a colorizer. Signal processors act on the signal after its initial generation and before its recording or display. The image or signal controllers often act on the processors; for example, a control voltage in the shape of a square wave may switch colors in a colorizer from red to green. Further, some processing devices, such as certain colorizers and computer-based systems, are more accurately signal/image generators since they operate without external inputs.

Controllers

Control over the signals which define image variables is central to electronic image processing. Whether manual or automated, it is exerted on a signal which defines an image and not on the image itself. A potentiometer offers manual control over voltage through the adjustment of a knob.

Control over signal parameters can also be automated rather than manual. The technique of voltage control allows the pots to be adjusted by another voltage rather than by hand.

Voltage control is the control of one voltage, often called the 'signal voltage', by another voltage, the 'control voltage'. If the control voltage frequency is within the range of human hearing, the signal can function both as a control voltage and as an audible sound; this dual role for signals and the resulting relationship between image and sound is a technique used frequently in electronic imaging. By use of control voltages, the problem of continuously varying changes is overcome; one can move between discrete values without having to proceed through intervening values. Control over digital and computer tools can be through knobs and switches or through computer programming.

The introduction of a computer to an imaging system like the one at ETC posed additional questions related to control and interfaces. At the ETC studio in 1977, where artists generally had week-long residencies, the software which ran the LSI-11 computer had to run invisibly, so artists did not need to learn programming languages. The computer in the ETC system functioned as an image generator and as an image controller:

Sherry Miller Hocking: A lot of the stuff that we did when we got involved in the computer which really started around 1976 had to do with Ralph's interest in trying to figure out interfaces, so that artists didn't have to become mathematicians and programmers and software experts – and artists could come to ETC and use the tools. We went through all kinds of variations from interfaces where you could turn the knobs to control the computer, to drawing tablets and digitizing tablets. Walter [Wright] and David [Jones] wrote software that would be transparent in the process, so that the artist didn't have to deal with the software but it was operating in the background. There was a very user-friendly interface to that software sitting on the screen that someone could easily use or work with. And that is the tactic now that we are using with the new stuff – Jitter and Max/MSP, trying to come up with these little modules that people can use.

Ralph Hocking: Yeah, but that's different. I think that when the Amiga was born I really got into the whole idea of what digital might be. So I spent many years playing with the Amiga. I firmly based my teaching on that machine and you had to learn how to connect to the language of that machine at a level that you knew what you were doing. I used Basic because machine language was too difficult for everybody else and too difficult for me. But we worked at a level of a language called the Director. They found all the possibilities that they could think of within that construct of the board and the CPU and all the little doodads that they had running beside it. They said 'OK. Here is what you can do visually, this is what will affect the graphic presentation on your screen'. And it was wonderful because you could write all kinds of really strange programs. You could write the programs yourself. You were right in there with the machine, you weren't dealing with Photoshop, or analogs, but working with machinery without having to be fed by technologists. (High 2006: 8)

At Synapse – the 1970s Syracuse video collective that spawned Bill Viola and where I taught early courses in video art – Carl Geiger purchased one of the first Altair computers. He generated non-objective stills by entering programs using nine flip switches: eight to define a byte, and one to enter it into RAM. There was no storage, so he had to enter your program each time you wanted to run it. Carl would run the computer output through the Synapse video switcher to play hell with the video synch pulses and keying voltages, and capture the output with a still camera. Occasionally we'd drag the school Moog Synthesizer from Franklin Morris's electronic music

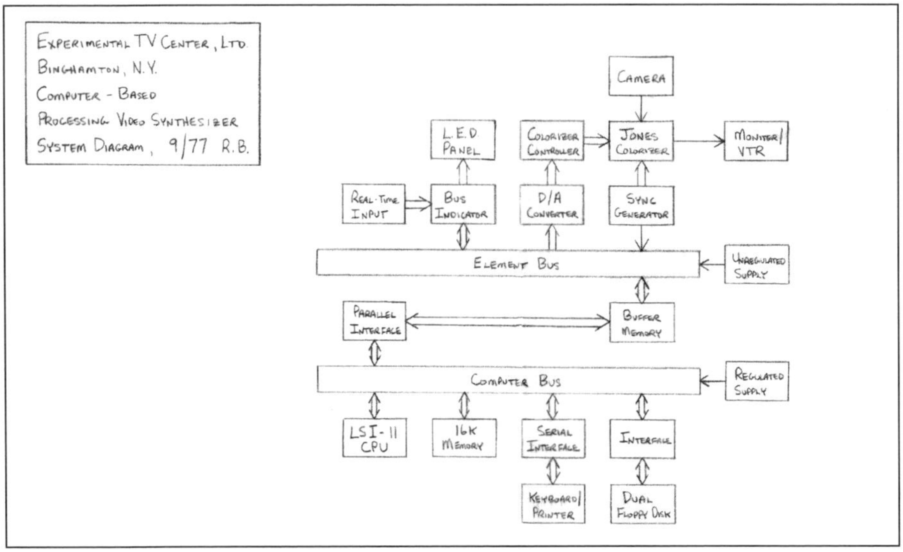

Figure 3. Signal flow documentation by Rich Brewster of LSI-11 computer project at ETC (1977).

classroom, where we'd try modulating the switcher and colorizers with the Moog output. (Edgar 2011: 3)

Processors

Signal processing – switching

Switching is a basic kind of signal processing from which other imaging techniques such as mixing, fading, superimposition, keying and wiping are derived. It turns signals on or off. Switching is the serial presentation of image or sound derived from two or more sources, and is achieved by the sequential change from one input signal to another. The term 'serial' can be misleading because as the rate of switching is increased beyond a certain frequency, we no longer see one image following another. Instead the image might appear to contain two or more source images simultaneously.

In general there are two kinds of switchers. In a 'broadcast' type, certain of the inputs are permanently connected to specific outputs. The result is efficient switching, but this type limits choices. In the 'distribution' type switcher, any input can be sent to any output. Because of the large number of cross-points, manual switching of this large number of points is time-consuming and potentially confusing. Most image-processing systems with distribution switching use a second system to electronically control these switching points.

Superimposition will appear if the rate of switching is faster than about 1/10 second, even though discrete images are presented. In this case, superimposition appears because the speed of the switching is such that the eye perceives not separate images in sequence, but images being mixed together.

Wipes can be thought of as switching within the frame. They appear as stationary or traveling horizontal or vertical divisions of the screen into two or more areas, each containing the corresponding portion of the input image.

This allows a spatial fracturing of the picture plane which is created by a rapid switching between or among images as they are constructed through time. It is this temporal aspect of switching which allows the spatial reorganization of imagery. Disparate images can be combined and possess technically logical relationships as a whole, yet call into play ideas of scale and our sense of 'real' time and space. These techniques as they have been employed in video have frequently been compared to Cubism, in their simultaneity of presentation of points of view, to the Suprematists with their use of purely geometric and often rectilinear formal properties, and to Surrealism, in their careful combination of logically dissimilar and impossible though seemingly realistic imagery. When live imagery is combined with prerecorded events simultaneously, our understanding of 'real' time and its apparent serial and progressive nature is challenged.

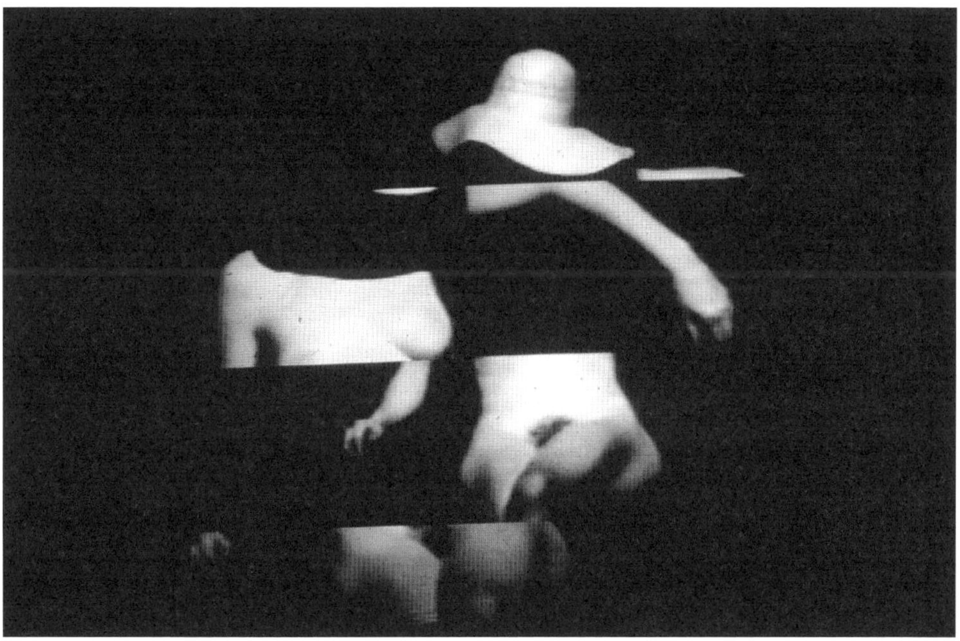

Figure 4. *The Hat* (Ralph Hocking, 1976). A combined use of oscillators and keying.

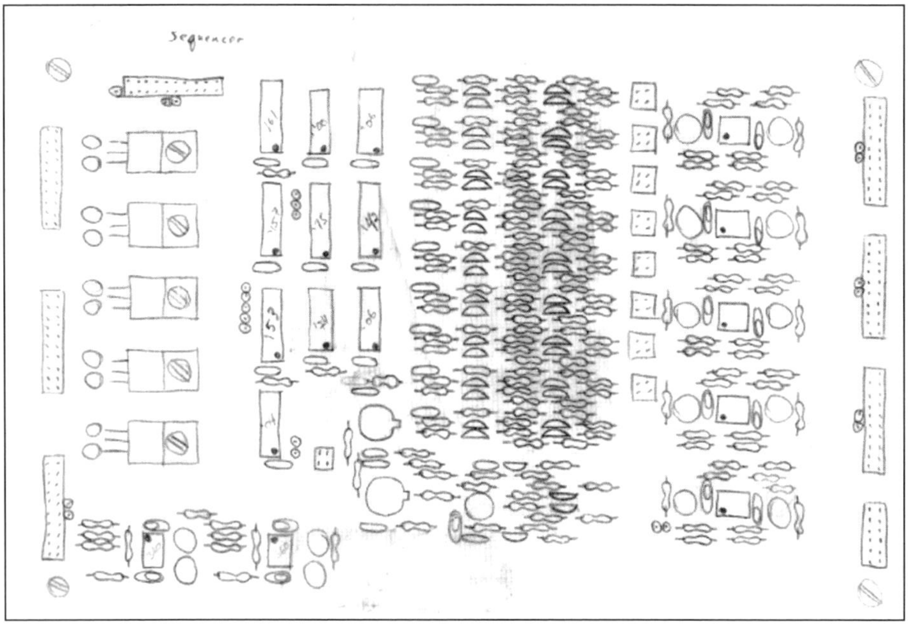

Figure 5. Schematic by Dave Jones for his sequencer.

Signal processing – sequencing

A sequencer may be thought of as a specialized kind of switcher. The switching series can be programmable so that the series can be easily repeated. The order is changeable so that the switching is not necessarily consecutive from input 1 through input 5, but may be preprogrammed so that you can jump from one to another in any order and repeat that order. Sequencers may contain a clock which controls the speed of the sequence. Switching began as a technique for cutting among a number of video images and thus allowed a number of points of view to be presented sequentially. Switching and sequencing are integral to video because they function within the timing limits of video imaging. As techniques of image processing, they are frequently used with 'real-time' images. In contrast, editing presumes that the images are recorded. If the switcher is able to accept videotape recorders as inputs, this sense of a 'real-time' sequence may be false, since 'recorded time' events are intercut with live events.

Signal processing – mixing

Mixers were developed in the 1950s as special cases of switchers. As with switchers, mixers used two inputs, but the images were combined in an additive way rather than an either/or way. In a mixer, the input image signals are added together and then, in effect, divided or scaled down so that the resulting signal output is within the prescribed limitations of signal amplitude of one volt, peak-to-peak. Mixing happens in real time

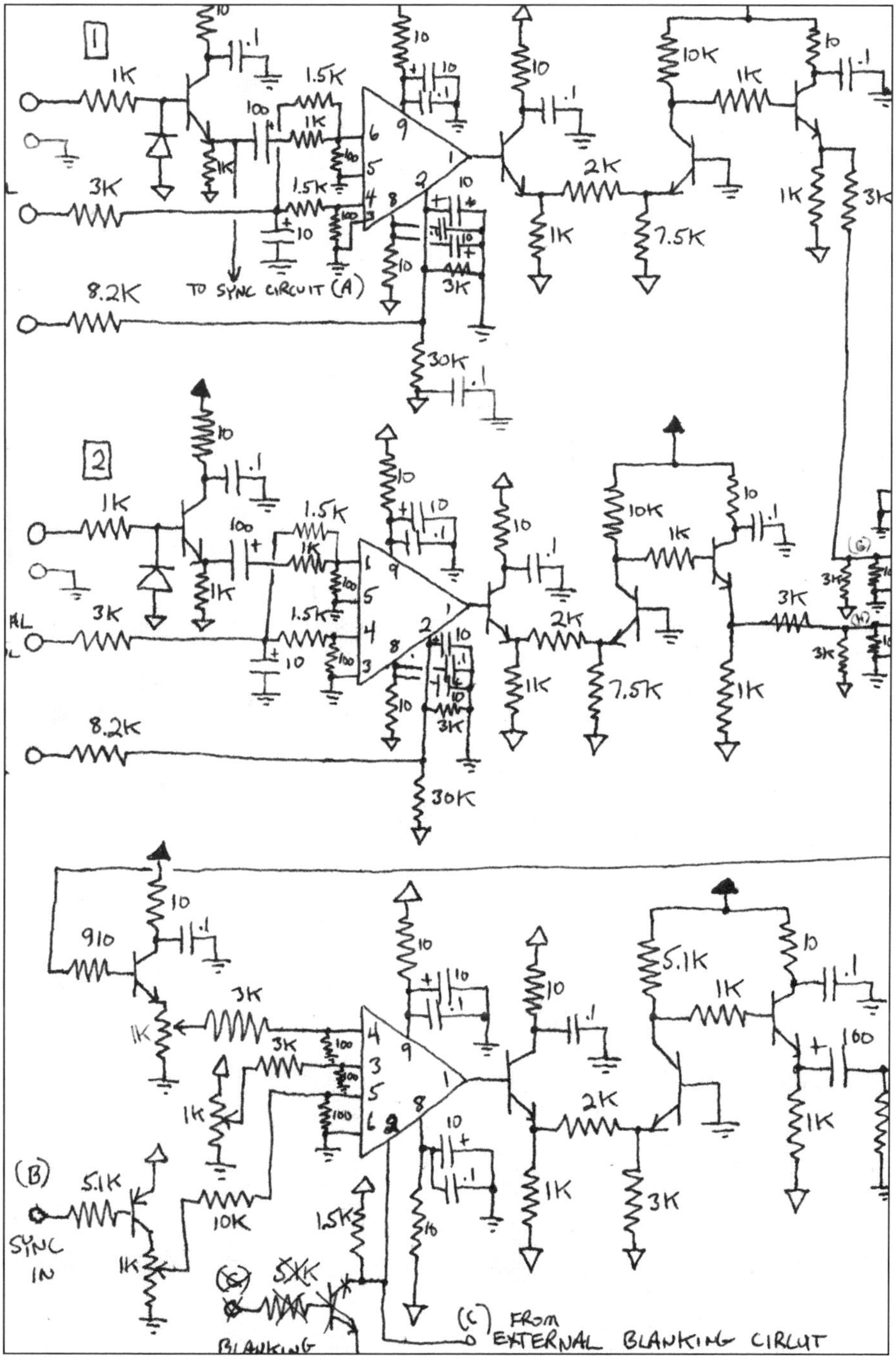

Figure 6. Rich Brewster's documentation of a Jones Keyer.

so that the output is the instantaneous sum of all the instantaneous amplitudes for each point along each horizontal line of the inputs. As the lines are built up into fields and frames, the resulting image is a mix of both source images. Mixers thus provide the means for the development of superimposition.

Signal processing – comparing/keying

Keying is another important imaging technique which is based on switching. It involves creating a window in an image so that corresponding portions of a second image appear in the hole. In monochromatic keying, the hole is defined by a gray value, therefore it will have different shapes depending on the spatial locations in the image which contains the specified gray value. The hole is created by a threshold voltage level which falls somewhere between 0 and 1 volt, the acceptable range for black-and-white video signals. As it comes in, the camera signal voltage is compared with this threshold voltage and at every point where the camera signal exceeds this threshold, the video is eliminated. As the video is eliminated, the keyer switches to a second input signal. In keying, the switching process occurs within the image rather than between 'whole' images. It occurs within each horizontal line rather than between fields or frames. It is then high-speed switching which occurs in real time, point by point, for each horizontal line. The clip sets a voltage level or gray value which determines when the keyer switches from A input to B input, thus inserting image B into the hole we created in image A. The clip can be a steady-state voltage which can be varied between 0 and 1 volt by adjusting a knob. The clip can be a signal from an oscillator, a sine, square or triangle waveform, or a complex waveform signal already constructed from other basic waveshapes. Or the clip can be a signal from one of the two video inputs or a third camera. A number of keyers can be banked or cascaded together; they can be operated independently of each other or the output of one can become an input for another. In this way, extremely complex, layered imagery can be constructed. Keyers frequently are voltage controllable; rather than increasing or decreasing the clip level by manually changing a knob setting, the clip level is changed by a signal voltage from another source.

Independent video technologists began to experiment with all these techniques in the early 1970s. Working with the Vasulkas, George Brown developed a Video Sequencer (also known as a Field Flip/Flop Switcher) with digital control in 1972. In 1973, Brown developed a Multi-keyer, an analog device with digital control (Spielmann 2004a).

In 1973, ETC technician Dave Jones modified an off-the-shelf Sony Special Effects Generator to accept direct sync interface with the Paik/Abe Video Synthesizer, with a provision for external wipe signal input. In 1974, Jones developed the Jones Colorizer, a four-channel voltage controllable colorizer with gray level keyers.

In 1975, at the ETC studio, a sequential switcher, along with the Jones Colorizer, were incorporated into the processing system. A 64-point push-button switching matrix was built and used in dance performances by the American Dance Asylum with Bill T. Jones and Arnie Zane (Hocking 2004: 9–10).

Processors – Scan Processors & Raster Manipulation (or Rescan) Units

> The concept in the Rutt/Etra is that the Rutt/Etra changes the time in which you see parts of the picture. It's a machine that manipulates images in time. I see it as a time processor. [...] I would allocate that feature as Steve's and my contribution. (Bill Etra in Vasulka 1978: 9)

Scan processors reorganize images by acting on the systems which control the scanning motion of the electron beam; these devices do not directly change the output signal but rather reorganize the way in which the signal is displayed. In a conventional CRT, the deflection systems maintain the horizontal and vertical orientation of the raster, producing a full and stable raster of light. Because the scan processor acts on the deflection system – the means of displaying the image – rather than the signal itself, the appearance of the image is changed; the signal which defines the image remains unchanged. In general, a scan processor consists of a CRT, the normal deflection system of which is modified; a set of controls over the deflection system which alter the raster in specific ways; and, finally, a rescan camera. Because the signal defining the image is not changed but only the appearance of the displayed image, the images must be reproduced optically; scan processors do not output a video signal which can be directly recorded. The reproduction of a scan-processed image is done by the technique of rescan: a video camera is pointed at the signal processor's CRT display and the image changes are rephotographed. The rescan camera output can then be recorded or used as an input source for further processing. Scan processors can operate on a variety of image or signal sources. Live camera or prerecorded information from tape or film, non-optically generated imagery derived from oscillators, and the output video signal from an image-processing system are all possible sources. Audio signals generated by microphones, radios or audio synthesizers may also be used. Scan processing can be performed on the raster itself, with no image displayed.

The types of image transformations possible with a scan processor include shape, position or placement, size, intensity and movement. The raster can be reversed along the X and Y axes by causing a reversal of the normal scanning operation. Reversal around the vertical dimension transposes right and left, creating mirror images, while reversal around the horizontal dimension transposes top to bottom, creating an inversion of the image. Collapse of the raster around the vertical axis produces a single horizontal line, while collapse around the horizontal axis produces one vertical line. Simultaneous collapse of both horizontal and vertical create a single, centered dot. The image can also be moved around the screen or repositioned and rotated in two-dimensional and apparent three-dimensional space. The deflection systems can also be controlled by external signals such as audio signals created by oscillators.

A brief history of scan processors

Ture Sjolander and Bror Wikstrom developed a modification to the raster in 1965–66, which was used in the artwork *Time* (Saask 2004). Other examples of raster-based tools are the Scanimate, the Rutt/Etra Scan Processor (also known as the Rutt/Etra Video Synthesizer) and Nam June Paik's Raster Manipulation Unit or Wobbulator.

The Scanimate was a programmable, computer-controlled animation system for primarily commercial applications. A high-resolution camera, usually looking at a high-contrast black-and-white film transparency, is input to the analog computer animation controller; this permits manual control over elements such as size, intensity, placement or rotation by generating varying voltages which are then applied to the horizontal and vertical deflection systems of a high-resolution monitor or scan converter. The changing voltages create different scanning patterns on the raster, and therefore different treatments of the image, displaying images which correspond to the voltage changes. The computer creates and stores these sets of voltages and then plays them back in real time on the rescan system; a second camera views the rescan monitor and its output is then

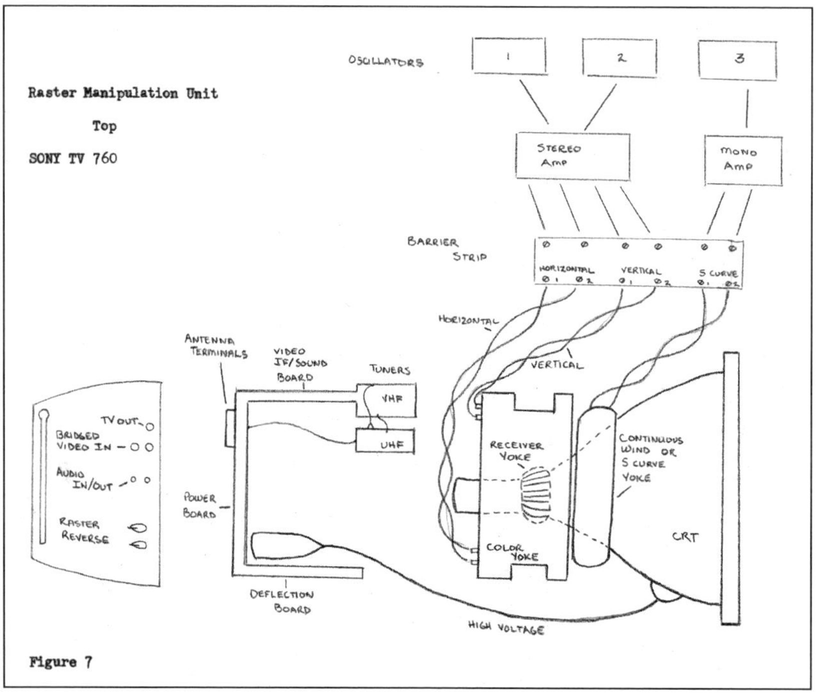

Figure 7. Block diagram of the ETC Raster Manipulation Unit
by Sherry Miller Hocking (1980).
See Color Plates 35 and 36.

fed to a colorizer. The colorizer operates on a set of five gray-scale values; each gray level has separate controls over red, green and blue. The output of the colorizer can then be recorded. The introduction of the computer to this system allows the changing voltages to be 'memorized' and stored and then played back in sequence to create movement in real time. The display was rescanned by high-end video camera or on film, and the output recorded. Walter Wright offers a cogent explanation of the working of the Scanimate in *Multimedia Performance: Beginnings* (2006).

Sid Washer also contributed to scan processing, making use of the deflection circuitry to generate patterns rather than the usual raster. These early prototypes often did not have video inputs because video cameras and oscillators at that time were prohibitively expensive. Initial experiments resulted in 1963 in a color television which displayed Lissajous patterns by connecting the deflection circuitry to amplifiers to modulate the red, green and blue beams separately (Burris 1978: 2). By 1965, he had developed a 3D color music device called the *Albatross*, which was designed as a performance instrument. The device used the modified color television, which was reflected in a vibrating mirror surface driven by a loudspeaker at a specific frequency; the patterns on the color set were controllable in terms of size and position, and provision was made for an external audio source, usually a phonograph, as an input. The vibrating reflecting surface created apparent three-dimensional effects. The system was essentially a pattern generator; images from this system were rescanned onto film. This device was exhibited, and a number were sold. Around 1972, Sid Washer developed a proposal in which he specified the parameters of the system, which was modular in design with the necessary signal patching. The system consisted of a display-processing unit, as well as control units, and generated a full-color display. The patterns generated could be controlled in real time in terms of width, height, brightness, color balance and spatial orientation; they were described by Washer as similar to 'computer-type graphics' generated without a computer. The system had provision for video inputs which he specified as character special effects and waveform generators, as well as audio sources and systems. In 1974 Washer began work with Steve Rutt and Bill Etra.

In California, Bill Hearn had worked since the mid-1960s on the development of a system for the production of color Lissajous patterns using phase-locked oscillators. Just prior to 1970, Hearn designed and built the Vidium, an XY display system which used a color CRT. The XY display was controlled through stereo microphones and the color was related to the speed of the electron beams. The red, green and blue guns were modulated by the velocity of the beam; each spectral color was assigned to a velocity and therefore changes in velocity produced color changes. Its primary use was in the direct generation of images from audio by electronic musicians, and it was never developed commercially. Although Hearn himself didn't consider the Vidium to be a video tool, since it relied on the functions of a conventional color television set, Hearn made significant contributions in the development of imaging systems (Vasulka 1978).

In 1969 Joe Weintraub exhibited *Audio-Controlled Television* at the 'TV as a Creative Medium' exhibition at the Howard Wise Gallery in New York. The work transformed 'music into a complex kinetic image on the screen of any color TV' (Howard Wise Gallery 1969). The volume of the audio controlled the brightness, while the pitch controlled color. The red, green and blue electron guns of the color receiver were governed by the frequencies of the audio divided into low, middle and high ranges. The patterns generated were dependent on the interaction of volume and pitch.

Nam June Paik is credited with the development of a Raster Manipulation Unit or 'Wobbulator', a prepared television which permits a wide variety of treatments to be performed on video images; this is accomplished by the addition of extra yokes to a conventional black-and-white receiver, and by the application of signals derived from audio or function generators on the yokes. The unit is a receiver modified for monitor capability; all of the distortions can thus be performed either on broadcast signals or, when the unit is used as a monitor, on images from a live or prerecorded source.

The distortions performed on the image result from the actions of audio signals on the yokes. Periodic and regular audio signals, such as sine or square waves, are often used when treating a video image; these signals are derived from an audio or function generator. However, any audio signal source may be employed; these devices include audio synthesizers as well as more conventional components, such as audio tape recorders, tuners, microphones or phonographs. These types of signals are most evident visually when used in conjunction with the horizontal or vertical collapse functions which reduce the raster to a horizontal or vertical line. These audio signals cause the line to distort in direct correspondence with changes in the audio signal.

In 1972, a raster scan manipulation device was constructed for the ETC studio and the construction as well as operation of the unit were documented in 1980. This how-to manual was distributed freely by ETC (Hocking, Brewster and Wright 1980) and is available online.[2]

In 1971, Bill Etra saw a Paik Raster Manipulation Unit at the Television Lab at WNET-TV in New York; he felt that because Paik's unit was AC-coupled it could only allow waveform distortion. Etra was interested in a modification to this unit that allowed for zooming and panning, the permanent placement of the image at any position within the raster. The Paik unit allowed for the placement of an image within the raster but it was not permanent. During 1972–73, with support from the New York State Council on the Arts through the TV Lab at WNET and with an investment of personal funds, the Rutt/Etra was prototyped. The Rutt/Etra was collaboratively designed by Bill Etra and Steve Rutt in association with Sid Washer, who assisted in the design and construction, and Greg Leopold, who was primarily responsible for the packaging of the system. In its initial stages of development, the Rutt/Etra scan processor was an oscilloscope using pots to change deflection voltages on the yokes. In the production model of the Rutt/Etra, the display CRT, which was rescanned, was available in 525- or 1050-line scan systems. Control was either manual or preprogrammable through the use of control voltages. The dual-trace production models of the system offered control over height, width, depth, shape, brightness, position and movement, as well as rotation in two- and apparent three-dimensional spaces. The height control on the display control unit

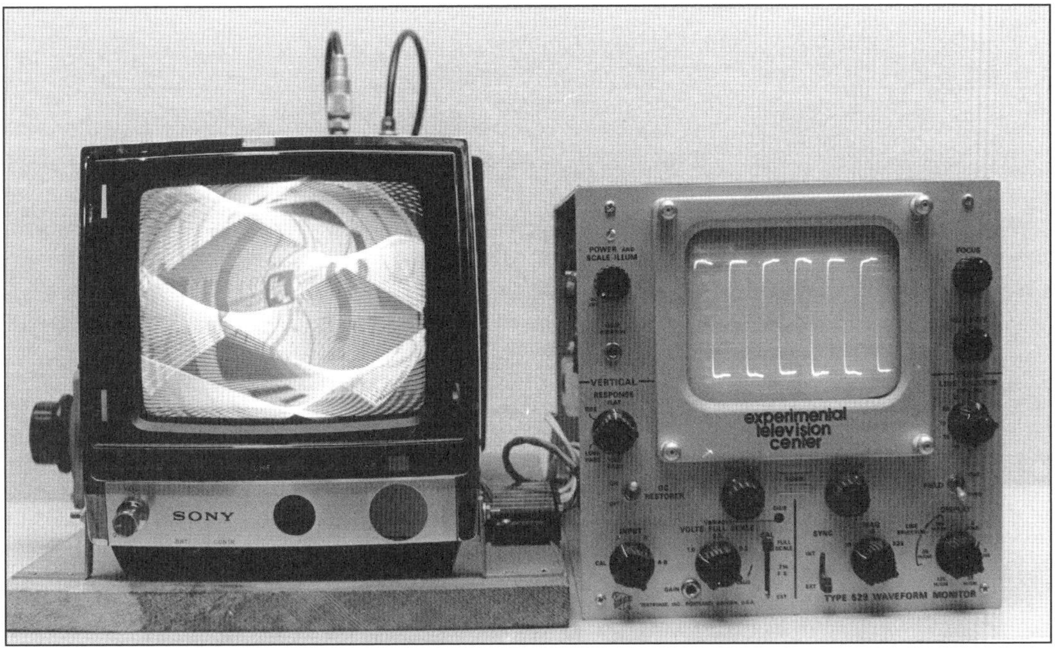

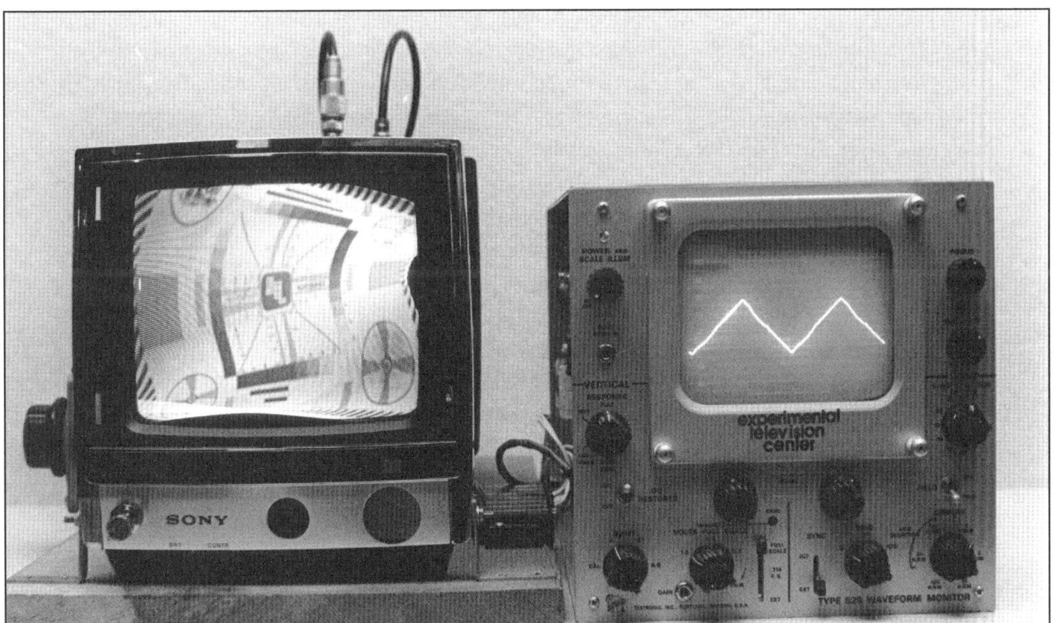

Figures 8 and 9. The Wobbulator showing the two different images resulting from two different waveforms modulating the scans of the same source image.

permitted reduction of the raster image to a line, expansion of the inverted image beneath and horizontal rotation. The width control allowed reduction, inversion and vertical rotation. Positioning control allowed vertical and horizontal placement, and horizontal and vertical axis adjustment was accomplished by a centering control. Variations of brightness were produced by an intensity control.

> Bill Etra: The concept in the Rutt/Etra is that the Rutt/Etra changes the time in which you see parts of the picture. It's a machine that manipulates images in time. In fact, I see them all as time processors. That's the way I see the whole scale of events. They are either parallel or sequentially time-processed images. (Vasulka, 1978)

Processor – colorizer

At the beginning of this discussion of video synthesizers, I called such devices the artist's paint and palette, and no module better fits this description than the colorizer. There are many designs for colorization devices both commercial and unique. In systems which use a camera encoder to generate the final output, the colors are determined by mixing red, green and blue components. While pleasantly reminiscent of mixing colored paint, this system is less efficient to use than the colorization made possible by video color parameters: luminance, chrominance and hue. With the latter system it is possible to pass a previously encoded color signal through the synthesizer and recover it unchanged at the other end (through the luminance channel). Now colors can be added by entering signals into the hue and chrominance channels. Where gray-scale encoding is used, such as in quantization, the single-hue parameter can produce rainbow-like effects. The Siegel colorizer, invented by Eric Siegel in 1969, uses gray-scale encoding to modulate all three parameters simultaneously. Dividing the inputs into three channels gives an increased degree of control over the final output. The commercially manufactured Grass Valley colorizer uses the luminance, chrominance, and hue parameters but is not voltage controlled and hence is not dynamic. (DeWitt 1976: 4)

The simplest type of colorizer is one that adds the color burst signals to the picture and modulates the phase of the subcarrier with the luminance of the black-and-white signal. These kinds of machines commonly have controls for chrominance (amounts of color), modulation (the amount of color change caused by the black-and-white signal), luminance (the amount of black-and-white signal mixed into the output), and tint or hue (the phase of the signal when phase shift initiated). This technique was used in the Riker industrial colorizer, George Brown's colorizer built for sale by C.T.L. Electronics in New York, and in the first colorizers designed by Bill Hearn for Electronic Associates of Berkeley, California. Similar effects can be achieved by using video signals as inputs to a standard RGB encoder (the machine that turns the signals from the guns of a color camera into color). If the same signal is used to drive all channels, and attenuators are present at the inputs, similar effects can be achieved. Video pioneers Nam June Paik and Shuya Abe used this technique in the colorizer section of their Paik/Abe Synthesizer.

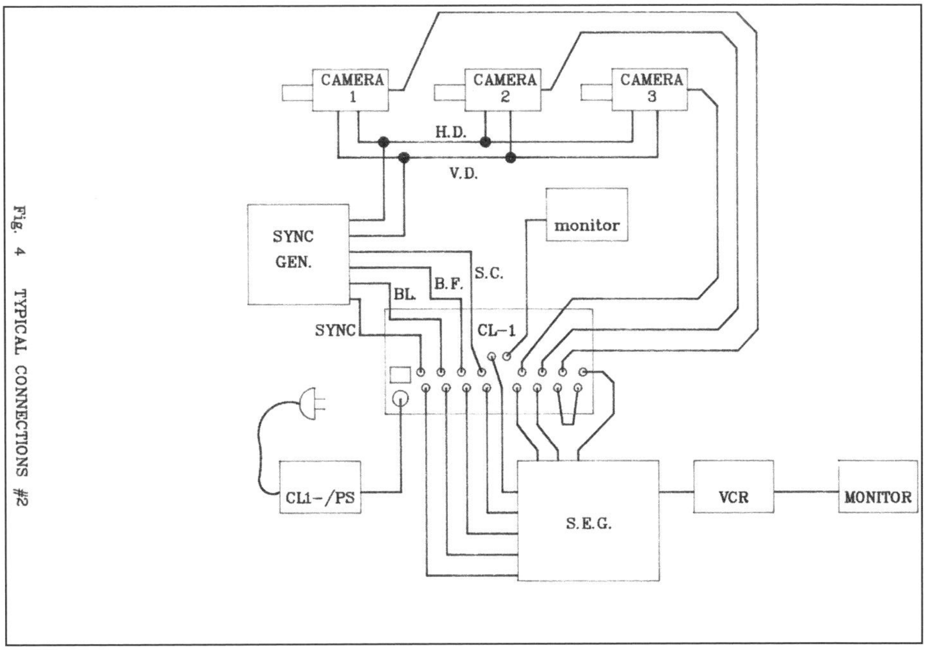

Figure 10. Jones Colorizer CL-1 Manual. Block diagram of multicamera operation. Documentation by Dave Jones and DesignLab (1986).

They added three more inputs (cyan, magenta and yellow), and allowed for different video signals into each input. They also added a phase shift on the whole system, which essentially changed the color value of each input.

The Image Processor of Chicago video artist Dan Sandin handles all signals as independent channels (Videography, December 1976) and, as a last step, puts them into an RGB encoder. Dan's machine goes several steps further, however, since it can process the signal to generate outlines and other complex effects before colorizing. One of the most interesting features of the IP is its ability to do amplitude classification, or quantization. Quantization is the division of the signal into a number of levels. Colors can then be added to each independent level. Most large video switchers use their keyers to quantize the picture, and by keying in color background generators, can colorize the image.

> This class of machines, direct video synthesizers, generates shapes by using analog or digital methods to switch color information from on or off, according to vertical and horizontal timings. In addition to the Sandin Image Processor, the EAB [Hearn] Videolab, the Colorado Video Colorizer, the Paik/Abe Synthesizer, and the Siegel Video Colorizers, other notable devices in this category include the Beck Video Weaver, the Chromaton [...]. (Etra 1979: 30)

> My second experience with knob twiddling was on the Paik/Abe Video Synthesizer (PAVS) – a much different 'beast' from the corporate Scanimate. It was an artists' machine – built by an artist for artists – more emotional and less intellectual. The PAVS was about color – not about counting, positioning and bending raster lines. Rather than the hard-edged, cartoon color of the Scanimate, it produced gorgeous, electronic watercolor. The PAVS had gain controls for the 7 video input channels. In addition each channel had a positive-negative toggle switch. There were controls for overall pedestal and gain and a large knob to affect the overall hue of the colorizer. (Wright 2006: 28–29)

In 1972, ETC began construction of the Paik/Abe Video Synthesizer, designed by Shuya Abe and Nam June Paik and built at the Center with David Jones and Robert Diamond, for eventual placement at the TV Lab at WNET TV.

> This system was used at the Center by Paik and the staff of WNET, including David Loxton and engineer John Godfrey, to produce a portion of Paik's *The Selling of New York*, a part of *Suite 212* broadcast in 1972 by WNET. Another PAVS was used early in the Center's Residency Program by artists such as Ernie Gehr, Hollis Frampton, Jackson MacLow and Nick Ray, and featured in the exhibition 'Work from the Experimental Television Center' at Everson Museum of Art, September 19 – October 1, 1972. In 1974 a set of oscillators were designed by Dave Jones for use as signal inputs to the synthesizer and as sonic modules (Hocking 2004: 9).

Other types of processing – modifications to camera and decks

As is discussed by Kathy High in 'Mods, Pods and Designs' elsewhere in this section, many early practitioners modified black-and-white cameras and decks to allow more control over the image or to create certain types of effects. Some of the modifications brought controls which were located inside the tool to the outside to allow manual adjustment by the operator.

The method of editing with the earliest of video decks involved physically splicing the tape.

The ½" open-reel video recording decks which appeared later were expensive, and the cheaper models did not allow you to make edits without interrupting the video signal, which caused glitches or short video erasure. Dave Jones speaks about modifications he made to reel-to-reel decks to assist with cleaner edits:

> There weren't any really hardcore editing decks that were in the low price range. There were decks that were considered editing decks, like the Panasonic 3130, which had the ability to do assembly edits, but did not have the ability to do insert edits. It

VSK-1 VIDEO SPLICING KIT

You can now make perfect videotape splices with no picture break-up or distortion using the SONY Videotape Splicing Kit. This kit, which can be used with any ½" SONY videotape, consists of a precision splicing block, a bottle of tape developer, video splicing tape, and a tape cutter. A pair of sanitary gloves is also included to prevent fingerprints and grease from contacting the oxide surface of the tape. (The tape developer has no adverse effects on the skin.) Read this instruction booklet carefully and save it for future reference.

1. Splicing block
2. Tape developer
3. Splicing Tape
4. Cutter
5. Sanitary gloves

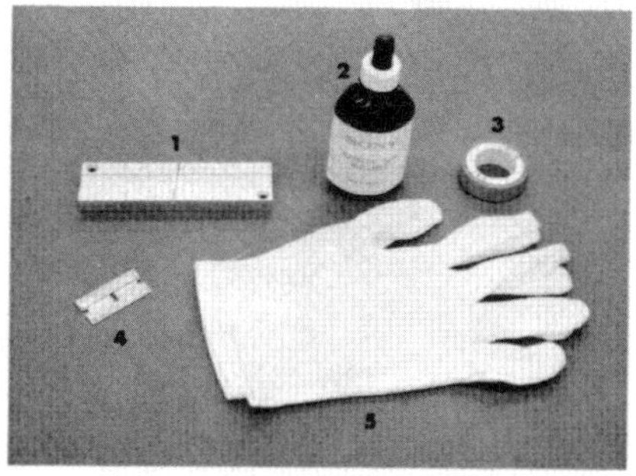

Fig. 1. VSK-1 Splicing Kit.

BEFORE SPLICING

It should be borne in mind that, due to the proximity of the audio and video heads, the audio and video signals are not recorded at the same point on the tape. (This is typical of all video tape recorders.) The audio signal is, in fact, recorded approximately 2½ inches ahead of the video signal. For all practical purposes, however, this difference in physical location may be ignored since, at a tape speed of 7½ ips, the difference in time between the two signals is only about 0.3 seconds.

What this means is that when a video portion is precisely cut from a tape, it may be missing the first 0.3 seconds of the corresponding audio.

The primary requirement of a good splice is continuity of control track pulses. Instructions are provided for developing the tape to observe these pulses. Portions of the tape that have been developed should be discarded. Do not develop any part of the tape that is to be played on the Videocorder.

It is most important that the tape and splicing block be kept free of developer, grease, or any other foreign matter to ensure a clean, noise-free splice. Use the sanitary gloves supplied when handling the tape.

Try a few practice splices on a section of unimportant tape to familiarize yourself with the procedure.

Figure 11. Sony VSK-1 Splicing Kit.

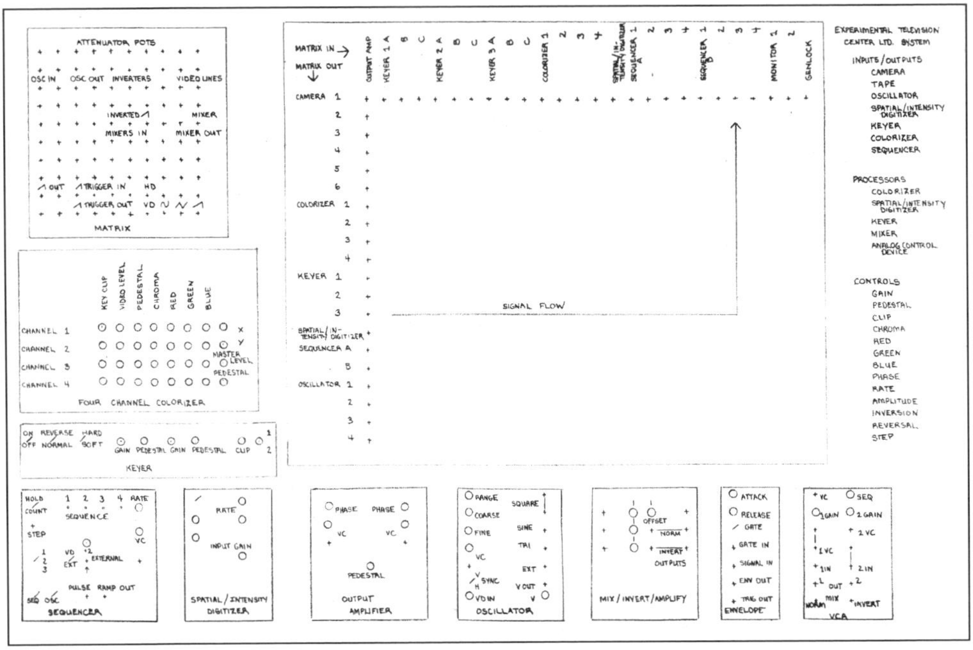

Figure 12. ETC Imaging System diagram showing the front panels of all the modules by Sherry Miller Hocking (c. 1980).

didn't really have a flying erase head that was needed to really do proper insert edits. And so I did a modification for that, where I put in a switch on a coil that would allow the main erase head for the tape to not erase the tape. And it was basically up to the video heads to kind of be strong enough to override the signal already on the tape, and you would be able to get an insert edit. Without that, when you went into the edit in that deck, you would end up cutting in fairly quickly; but when you cut out, you would have maybe eight seconds of noise before your original image came back, because the erase head was just a bulk erase head and it erased everything on the tape, as long as the button was down to go into record. (Jones 2007)

Other types of processing – recording and playback

Time-delay playback

This involved placing two ½" open-reel playback decks a short distance from each other. The tape was threaded onto the first, which would be the recorder. The tape continued on to the second deck, which would play back the images recorded by the first deck. The space in between the decks resulted in a noticeable time delay, since the recorded information didn't reach the playback head for the duration of the separation.

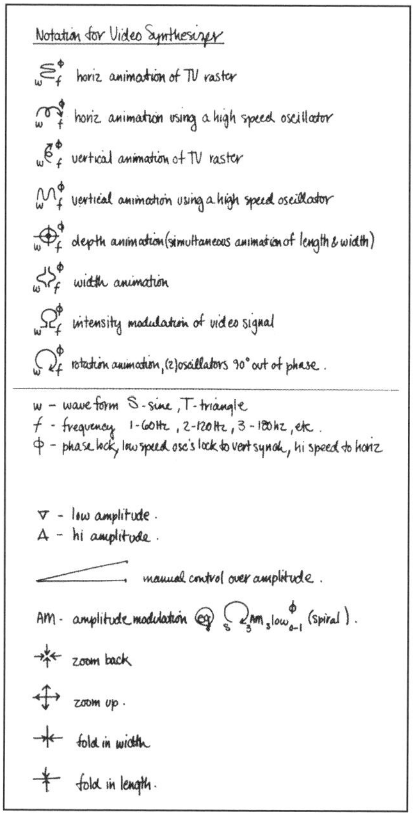

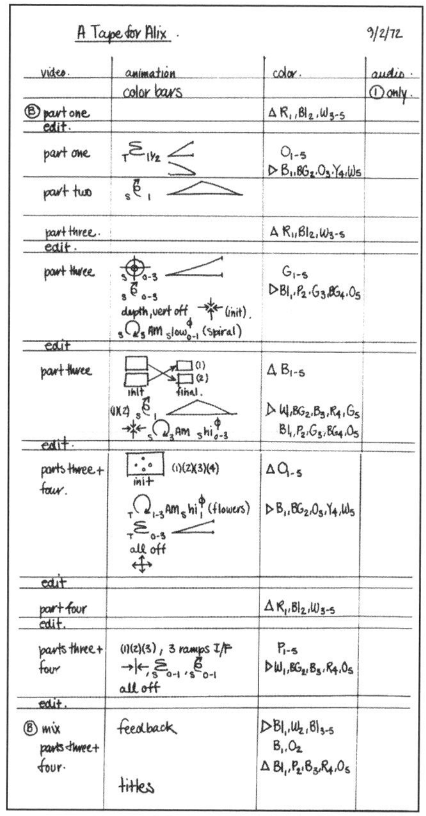

Figures 13 & 14. Walter Wright notation system and a tape score.
From program notes, The Kitchen, New York City (1972).

Image degradation

This could be achieved by making a recording and then playing it back while a camera was recording the image from the monitor. As this was repeated, the image took on textural qualities.

Sync – drift and roll

With this technique, artists could explore the space of the video image, setting the image drifting across the monitor screen or from one monitor to the next by controlling the horizontal sync. If the vertical sync was off, the result would be an image that rolls from bottom to top. Joan Jonas's *Vertical Roll* (1972) is a well-known example. Peer Bode also explored this technique:

> *Video Locomotion (man performing forward hand leap)* (1978). Homage to Eadweard Muybridge. Muybridge photo grid put into a video system space. Movement is created by detuning the horizontal and vertical video synchronization (time base) signal.

> Drift and horizontal doubling takes place. When the horizontal frequency doubles the man doubles on top of himself. A para-cinema shutter discovered by combining video luminance keying of the sync signal together with time base drifting. A basic video structure; sync and a basic video process; keying together create a video based film like shutter, simulating a crude persistence of vision system. Primitive physical structures of the video signal, combined, bring Muybridge's photo grid into an electronic animation of false movements. The overall time composition is a simple structure of increased time display of the photogrid and decreased time duration of the photogrid drifting. Two b&w cameras, time base detuning and keying. (Bode 2001)

The Vasulkas have used drift in many of their early works including *Home* (1973) and *Golden Voyage* (1973). George Brown constructed a process for them in 1972, which the Vasulkas called the Horizontal Drift Variable Clock which output the sync which controlled the speed of the horizontal drift (Spielmann 2004b).

> Using the colorizer, multi-keyer, and switcher, as well as horizontal drift, *Home* consists of three sequences in which still lifes are set in motion – e.g., an apple drifting past a teapot on a kitchen stove. *Golden Voyage* refers directly to Magritte. It is a sort of animation of his painting *The Golden Legend*. 'We were looking at this picture and we were joking about how many cameras we'd need to reproduce it,' Steina explained. 'Of course, three. One camera would be on.' These images were combined using the multi-keyer and set in motion via horizontal drift. Loaves of French bread embark on a journey. They travel across various backgrounds - a mesa, a beach, a building as well as a reclining nude woman. Initially mere loaves, the breads take on phallic connotations as they encircle the woman - an attempt at absurdist humor. (Vasulka and Vasulka nd: 2)[3]

Feedback

Feedback results when the camera is pointed at a monitor which is displaying the image from the camera. There are many variables, including lens and monitor adjustments, ambient lighting, and angular orientation of the camera relative to the monitor, each affecting the resulting feedback. By pointing a camera at a monitor, you are electronically 'rephotographing' or rescanning the images on the monitor screen. This image can be a live image from another camera or the output of a processing system, a prerecorded image from videotape, or, if the monitor is displaying the camera which is pointing at it, feedback. The output of the feedback loop can then be further enhanced using other processes such as keying or colorization:

> Video technology moves visual information from here to there, from camera to TV monitor. What happens, though, if a video camera looks at *its* monitor? The information

> no longer goes from here to there, but rather round and round the camera-monitor loop. That is video feedback. From this dynamical flow of information some truly startling and beautiful images emerge. (Crutchfield 1984: 191)

> 'Feedback' in this usage is a technical term, designating the procedure of connecting camera and display-monitor in a loop, the camera photographing the display and feeding the result back into the same display. [...] A feedback image is not a picture of anything, finally; it is a balance of purely electronic forces below the threshold of perception. It is our entrance into that very specialized branch of video called image synthesis, in which images are not records but creations achieved by manipulating the basic electronic forces at work in video cameras and displays. (Arn 1973: 20–21)

As Bill Gwin explained:

> Feedback is the image configuration most video experimenters discover first. It is produced by the most simple complement of electronic tools, a camera and a display monitor. By manipulating these two objects the artist can conjure limitless variations of stunningly complex imagery. In the early days of discovery, feedback is magic: spirals, flowers, mandalas burst forth with the touch of a fingertip and regenerate themselves indefinitely on the screen. Later, for some, feedback's simplicity becomes deceptive and its ease occasions serious questions of composition. [...] Video feedback is produced by aiming a camera at a monitor; the camera actually takes a picture of itself. The patterns thus engendered can be altered in several ways, by exerting various controls over the electronics, and by affecting the optical path of the picture/monitor loop. (Gwin n.d.)

The term *feedback* was also widely used by artists and designers in discussions of systems theory. Phil Morton and Dan Sandin used it as a descriptor for their Image Processor, along with the concept of interaction. They stressed the importance of the machine providing real-time visual and auditory information to the artist so that he or she could interact with the system. If the machine cannot provide complex real-time output, then the system must include predictive tools to allow the artist to 'guess' what is going to happen:

> SO I WENT ABOUT DESIGNING A VIDEO IMAGE PROCESSOR [...]. The primary idea was to do instant modification of instant feedback [...] simply the experience of getting INSTANT feedback is an extraordinary thing [...] it is a NEW thing with the realization of electronics [...]. (Sandin and Morton 1973)

Modularity, patching and notation

A video synthesizer or 'image processor' is a general term referring to an assemblage of individual modules which produce video or control signals, or process signals, all of which are then integrated into a single system. Arn and others have noted the distinction between direct synthesizers – those which allowed for no inputs – and indirect systems which alter the signals from a variety of external input devices. In reality, systems like the Vasulkas, the Hearn, the Sandin IP and the ETC system permit a variety of camera-based and electronically generated inputs that can be interconnected in different ways. Often, patch cables were used to physically determine the paths of the signals as they moved through modules. Later, as systems became more complex, designers used a matrix which was essentially an XY configuration of inputs and outputs. By positioning a slider, for example, the connection was made between an input and an output. The specific architecture of the system changed as modules were added to or removed from the patch or the system.

> Modular design is essential in video, because it permits parallel and simultaneous processing of high-frequency signals [...]. As in the development of the audio synthesizer, artists have provided engineers with functional module building blocks which efficiently accomplish commonly needed functions. Modular design also permits a wide range of interconnections depending on the 'patch' made between them. For example, a system of only 8 modules, each with a single input and a single output can be patched in over 40,000 different ways [...]. While modular systems provide both variety and efficiency, they also can present the artist with a confusing welter of two-ended wires which makes live performance difficult and leaves him with no permanent record of his patch. The first step in improving this situation came with the introduction of the matrix switching systems of the Arp and EMS synthesizers, adapted for video by Woody and Steina Vasulka. These systems have manually set crosspoints and permit patchfields to be recorded by graphic notation. Going a step further Don Buchla and Bell Labs have developed computer controlled patchfields which are notated with a verbal language. (DeWitt 1976: 2)

Many artists developed methods for documenting the pathway of the signal through various modules as they were interconnected, in part so that a particular effect could be recreated. In 1972, Walter Wright created a notation system, which he described as recording 'the state of the machine'. He used the system to create visual scores for his video compositions. He first developed this for the Scanimate, but later applied it to tools at the Experimental TV Center. Because the system was constantly changing, he modified the notation system to one which documented patches (Wright 2006).

Image-processing system

The System at the ETC studio was designed to be modular. Each artist could select subsets of sources, processors and controllers, and 'build' an individualized system. Artists applied these electronic processing tools within a variety of formal structures, ranging from narrative and documentary to abstract, formalist and didactic. The works were displayed in diverse ways – single channel, multichannel, live, interactive, sculptural or performative. Artists such as Woody Vasulka and Peer Bode described their works as documents of specific processes. The process could be scored, as for example by Walter Wright, and a tape specifically created from that score. Processed imagery could also be employed to convey or evoke emotional or spiritual states as in the works of Barbara Buckner, or to unfold a complex narrative as in works by Irit Batsry, Alexander Hahn or David Blair. They were intimately tied to music in the works of Reynold Weidenaar, or electronic sounds as in the early works of Gary Hill or Kjell Bjorgeengen.

The work could be performative, such as the works of Peer Bode with Meryl Blackman and the American Dance Asylum, with Arnie Zane, Bill T. Jones and Lois Welk. *Couple 513* (Everson Museum of Art, 1976) and *Movements for Video Dance and Music* (Herbert F. Johnson Museum, Ithaca, 1976) were both early explorations of video/dance. *Synergism* was a travelling performance collaboration between Woodstock Community Video and the Experimental Television Center (1975 and 1976). Participants varied but included Tobe Carey, Gary Hill, dancer Sara Cook, Ken Marsh and Walter Wright. *Ballet Di-Gi-Tal*, a live video and dance performance by Connie Coleman and Alan Powell with the Dada Processors, was presented in 1989 at International House in Philadelphia, and also at CEPA Gallery, in 1992–93.

The video system could also be set up by selecting signal sources and processes and then run as a sculptural installation, or set to respond to aspects of the outside environment using triggers. *Cloud Music* was a mixed-media installation created by David Behrman, Bob Diamond and Robert Watts and exhibited between 1974 and 1979. The installation consisted of a black-and-white camera which was pointed at the sky. A video analyzer allowed the user to position the placement of six targets on the monitor of the sky image. The device generated six control voltages which were each proportional to the light value of the associated target. As the light values of the passing clouds changed when moving across the crosshairs, the voltages changed. These varying voltages were sent to an electronic music system, which converted the signal changes to varying tones. *Cloud Music* was exhibited first at the Electric Gallery in Toronto (1974), and at the Experimental Television Center in Binghamton (1976), as well as other venues.

The ETC system continually evolved. As new tools were prototyped, different control methods were designed, and as consumer equipment finally began to catch up with independent video's early years, the building blocks would change and artists continually created new systems.

References

Arn, Robert (1973), 'The Form and Sense of Video', *Artscanada: The Issue of Video Art*, 30: 4, pp. 15–21.

Bode, Peer (2001), 'Biography', *Experimental TV Center* website, http://www.experimentaltvcenter.org/peer-bode. Accessed July 17, 2012.

Buckner, Barbara (1978), 'Light and Darkness in the Electronic Landscape: Some Aspects of the Video Image', Experimental TV Center website, http://www.experimentaltvcenter.org/light-and-darkness-electronic-landscape-some-aspects-video-image. Accessed July 5, 2012.

Burris, Jon (1978), 'Interview with Sid Washer' [unpublished], Owego, NY: Experimental Television Center Archives.

Crutchfield, J. P. (1984), 'Space-Time Dynamics in Video Feedback', in David Dunn (ed.) (1992), *Eigenwelt der Apparatewelt: Pioneers of Electronic Art*, Linz, Austria: Ars Electronica, pp. 191–207.

DeWitt, Tom (1976), *Design Device – A Unique Video Synthesizer*, Owego, NY: Experimental Television Center Archives, pp. 1–16.

Dunn, David (ed.) (1992), *Eigenwelt der Apparatewelt: Pioneers of Electronic Art*, Linz, Austria: Ars Electronica.

Edgar, Robert (2011), 'A Context for First-Generation PC Art: Times and Aesthetics in the Early Days of the Personal Computer: 1970s – 1980s', http://www.robertedgar.com/wp-content/uploads/2011/11/RBE_LEA_Context4FirstGenPCArtists.pdf. Accessed July 16, 2012.

Etra, Bill (1979), 'Colorizers Come of Age', *Videography*, 4: 4, pp. 30–31.

Frampton, Hollis (1977), 'The Withering Away of the State of the Art', in Douglas Davis and Allison Simmons (eds), *The New Television: A Public/Private Art Essays, Statement and Videotapes Based on 'Open Circuits: An International Conference on the Future of Television'*, Cambridge MA & London: MIT Press, pp. 24–35.

Furlong, Lucinda (1983), 'Notes Toward a History of Image-Processed Video: Steina and Woody Vasulka', *Afterimage*, 11: 5, pp. 12–17.

Gwin, William (n.d.), 'Video Feedback: How to Make it; An Artist's Comments on its Use; A Systems Approach', *National Center for Experimental in Television Reports 3*, Owego, NY: Experimental Television Center Archives.

Hagen, Charles and Kite, Laddy (1978), 'A Syntax of Binary Images: An Interview With Woody Vasulka', *Afterimage*, 6: 1–2, pp. 20–31.

High, Kathy (2006), 'Kathy High Interviews Ralph Hocking', April 16, Owego, NY: Experimental Television Center Archives.

Hocking, Sherry Miller (2004), 'The Evolution of Thinking Machines', *The Squealer*, 15: 1, pp. 9–10.

Hocking, Sherry Miller, Brewster, Rich and Wright, Walter (1980), 'Raster Manipulation Unit: Operation and Construction', http://www.experimentaltvcenter.org/raster-manipulation-unit-operation-and-construction. Accessed July 10, 2012.

Hocking, Sherry Miller, Brewster, Rich, Jones, Dave and Rudolph, Hank (1974–86), 'Image Processing Manual: Studio Manual for the Experimental Television Center' [unpublished], Owego, NY: Experimental Television Center Archives.

Howard Wise Gallery (1969), *TV as a Creative Medium* [exhibition brochure], New York: Howard Wise Gallery.

Jones, Dave (2007), 'Interview with Mona Jimenez and Kathy High', July 31, Owego, NY: Experimental Television Center Archives, pp. 10–13.

Saask, A. (2004), 'The Artist that invented Computer Animation', http://turesjolander.homestead.com/The_Artist_that_Invented_Computer_Animation.htm. Accessed July 12, 2012.

Sandin, Dan and Morton, Phil (1973), 'Conversational Process: Feedback Loops' [unpublished], Owego, NY: Experimental Television Center Archives.

Spielmann, Yvonne (2004), 'Video and Computer The Aesthetics of Steina and Woody Vasulka: George Brown', The Daniel Langlois Foundation, http://www.fondation-langlois.org/html/e/page.php?N004. umPage=455. Accessed July 15, 2012.

—— (2004b), 'Video and Computer The Aesthetics of Steina and Woody Vasulka: Horizontal Drift Clock', The Daniel Langlois Foundation, http://www.fondation-langlois.org/html/e/page.php?NumPage=441. Accessed June 12, 2012.

Vasulka, Woody (1978), 'Bill Etra and Louise Etra in conversation with Woody Vasulka' [unpublished], Owego, NY: Experimental Television Center Archives.

Vasulka, Woody and Nygren, Scott (1975), 'Didactic Video: Organizational Models of the Electronic Image', *Afterimage*, 3: 4, pp. 9–13, http://www.vasulka.org/archive/4-30c/AfterImageOct75%285024%29.pdf. Accessed July 17, 2012.

Vasulka, Woody and Vasulka, Steina (n.d.), 'Drift! The Vasulkas', *Art and Science Laboratory*, 1–8, http://www.vasulka.org/archive/Publications/FormattedPublications/DRIFT.pdf. Accessed July 17, 2012.

Wright, Walter (2006), 'Multimedia Performance: Beginnings', http://www.experimentaltvcenter.org/sites/default/files/history/pdf/wrightmultimedia_2599.pdf.

Notes

1. See Hocking and Brewster (1974–86). These manuals were used by the artists in the studio and in workshops. They were also used by other media centers, for example Media Study/Buffalo.
2. See Hocking (1980), http://www.experimentaltvcenter.org/raster-manipulation-unit-operation-and-construction
3. Available at http://www.vasulka.org/archive/Publications/FormattedPublications/DRIFT.pdf . The Vasulka document appears to be an earlier version of Lucinda Furlong's (1983), 'Notes Toward a History of Image-Processed Video: Steina and Woody Vasulka', *Afterimage*, 11: 5 (December), pp. 12–17.

ETC's System

Hank Rudolph

The text below was originally an introduction to a lecture on the ETC studio system in May 2005 as part of the Experimental Television Center International Student Residency. The ISR was a college-level course offered through Alfred University and co-taught by me and its director, Pamela Hawkins.

The two-week summer program ran for seven years, from 2000–06, and as an unaccredited workshop for four years prior to this. It was conducted at the ETC studio in Owego, NY, with students participating from several schools, including Atlanta College of Art, the University of Buffalo, the Rhode Island School of Design, the University of Georgia, Syracuse University and Alfred. For many of these years, a team of media artists, including Aaron Miller, Monica Duncan, Matt Underwood and Annie Langdon, offered additional instruction to, and often collaborative productions with, the students. Aaron taught the Max/MSP/Jitter workshops, which he integrated into the ETC studio workflow. Carolyn Tennant transcribed the original talk, which I later revised for clarity.

The text does not attempt to address the history of the tools in the studio or interpret the intentions of the designers, but rather to offer a conceptual overview of the system as a whole, and its defining characteristics as a unique mode of working in the time-based visual arts. This view is based in part on my own observations of artists using the studio in my years as ETC program coordinator, as well as my own personal experiences with the system.

I knew that many of the students were familiar with Final Cut Pro and Adobe After Effects. Using this as a starting point, I wanted to pose the question of why one would use such a system as ETC's, and what were the parallels to, and departures from, the approaches they already knew.

Now that the ETC studio has closed its doors, and the issues of materiality and process in video have shape-shifted with high-definition, a larger goal in revising this is to offer a blueprint for future modes of working and to address what types of studios, collaborations and tool sets (both hardware and software) have since happened and may yet still be needed.

Introduction

Before I go into the technical aspects of the system, I'd like to give a brief conceptual introduction to the ETC environment. I wanted to start by asking: Why would one want to work this way? Why use this type of studio? This model of working is not for everybody. What type of information will you get out of this residency that you can apply elsewhere? Is this workshop going to be only a long list of arcane instructions for operating a set of machines that you'll never encounter again? In addition to a body of work, I'm hoping that you come out of here with a way of rethinking the moving image, and that you apply that new perspective to whatever tool sets you use or even invent through software or circuit-bending.

There are many aesthetic and philosophical approaches to working with time-based media, but there are basically three technical approaches. Film production is the first. The second is the type of post-production software that many of you have worked with, such as Final Cut Pro, Pro Tools and After Effects. The third area is the one represented by the tools behind me. These tools have never had an adequate label. I'm still searching for one. Collectively the technology has been called 'video synthesis' and 'image processing' at various times.

I don't think it's productive to pit these approaches against one another or to decide which is the best way to work. Even if you never wind up using film, I hope you'll eventually learn about the process of filmmaking, and study the history of experimental film practices. Knowing more about each one of these areas can inform how you approach the others. Each field of study helps one develop a concept of raw material; each offers its own unique technological systems for working with that material, and each of those systems involves the artist in a markedly creative process.

In making distinctions and comparisons among these ways of working, I don't necessarily want to create a dichotomy between 'analog' and 'digital', because I think such a division obscures the real issue. That issue has to do with the underlying characteristics of the system as a whole, and how those characteristics add up to a radically different creative dynamic for the user. While most of these devices are in fact analog electronic tools, some, such as the frame buffer, are more of a hybrid digital and analog technology. In addition, certain programs such as Max/MSP/Jitter, have far more in common with the features of ETC's analog system than they do with post-production software.

I'm discussing the system as a whole, and taking a snapshot of where it is now, comparing it to other systems that exist now, rather than discussing how it got here. I'm not going into the very important context of the history of the development of the individual tools. This could be considered a synchronic rather than diachronic perspective.

When the TV Center first came about in 1971, there was of course film, but there were no digital video tools and the existing analog ones were mostly diluted consumer versions of those used in the television industry, notably the video switcher, or special effects generator. We have a few of those in the system.

Not only were there no blueprints for building individual machines, there was no concept of the personal studio as an electronic image-making environment in the artists' vernacular. There was just a sense of invention from the individual designers and from the artists with whom they communicated or collaborated. First and foremost, the ETC environment was an early model for what a personal electronic studio might look like. Now people have personal studios for electronic image-making, although clearly they are not at all based on this environment. The electronic studio that we know of today is a very different construct. It can be the laptop, camcorder and external hard drive, something very compact and even portable, and expandable mostly within the domain of the software. The environment you see here is much messier, more sprawling, sculptural and open-ended. This particular model of a studio never entered the vocabulary of the rapidly expanding media artist population, and the tool sets here never wound up proliferating in the consumer marketplace. Designed out of the natural curiosity of creative engineers, with hopes of eventually becoming more commonplace, these machines instead continued to be rare and eventually took on the status of 'precious'. Over the years I feel as if the image of the ETC studio somehow shifted from a paradigm of a personal artist's studio into a sort of hands-on museum of obscure audio-video devices. This was not the original direction.

I'd like to offer an overview of what to expect from this workshop. I like making lists. I'm going to outline six fundamental characteristics of the ETC environment, look at them individually, and compare them to the other two approaches of time-based image making: film and post-production software.

The six characteristics are:

Real time
Open-ended architecture
Indeterminacy
Interactivity
Sound-image synchronization
Electronically-generated, or camera-less images

Real time

Real time can have many meanings, but in this case it simply means that the artist can immediately witness the results of parameters being changed with knobs, switches or other external controls. It can also mean that some of the source material, such as camera images, can be live.

Real time does not have some absolute value. It only has meaning relative to the body, or to our experience or perception of the electronic events unfolding. Those events are not actually simultaneous. They're causal. They're happening in sequence but very quickly. It takes time for the signal to travel from source to destination, or from a studio camera to a display. It takes time to make changes to that signal. But to us, they have the illusion of simultaneity.

Why would that experience be useful in the creative process? The significance of real time is apparent for live performance and also in interactive installations, both of which are site-specific. The audience has to be there. Why does the artist have to be here to produce something that will result in a single channel prerecorded work to be experienced by an audience at some later time? How does working in real time contribute to that final piece? Are different modes of working, or processes, relevant in the reception of the final product? When you are sitting there, caught up in the unfolding experience of a completed time-based work, trying to make sense and meaning of layered images, does it matter whether those images were compiled through laborious filmic techniques in optical printing, or rendered with key frames in After Effects, or performed as a continuous take in a dynamic electronic patch?

It depends upon the type of work and the artist's intentions. In some cases it does and often it does not. But does this imply that the system is designed to produce only certain types of work? On the contrary, a large variety of works are made here. The productions of those works share a common creative dynamic, the most important aspect of which is the spontaneity offered by these real-time systems. I think that it radically alters the decision-making process. It adds a performative element to image making.

Working in this studio is like putting yourself in an installation, and then setting the parameters of that installation. There are certain sets of choices: the combination of image sources, some live, some not; the layering of those images; the length of recording time; and the changes that will take place over that time – choices that do not fall neatly into the categories of production or postproduction. You have a recorded block of continuous time, like a document of a performance. Next you must decide what to do with this block of time. The process of getting from that experience to a finished piece can entail a whole other set of editing decisions. Are you faithful to the idea of a performance, editing very little if at all? Or do you chop up what you have, resulting in a work completely out of the context of the original performance and taking the images in a very different direction? Artists like Andrew Deutsch and Matthew Schlanger stay within the integrity of that performance, and edit very little. That might not be your approach, nor does it have to be. But working here will raise these questions and likely result in a final product that you would not have made otherwise using other processes such as rendering or compositing. (See Color Plate 44.)

Open-ended architecture

Open-ended architecture is another important characteristic of this studio. Rather than thinking of the system as delivering a series of effects, consider how it works as a whole. How can you break it down into some very simple discrete ideas and then build these back up into larger structures with many combinatorial ideas available? In analog electronic hardware, the most useful model for this has been modularity, and in software, perhaps it has been the visual programming languages of Max/MSP/Jitter and Processing.

Aesthetic assumptions are built into every piece of hardware and software we use to make art, and often tools are packaged to presuppose certain applications, perhaps even designed as metaphors or emulations of other existing systems.

Many of the television industry's early techniques for combining images were developed as electronic emulations of filmic processes: keying for matting, mixing for double exposure, and so forth. These were built into hardware designs that best suited the formulaic applications of broadcasting. When consumer video technology was first marketed in the late 1960s and early 1970s, the few tools available were scaled-down versions of these broadcast devices, called the video switcher or special effects generator (SEG). A defining characteristic of the SEG was that these techniques for collage and montage were mutually exclusive options. You either keyed between image A and B or you mixed, or switched, or used a wipe pattern.

An altogether different technology with completely distinct aesthetic concerns was developing in parallel to the consumer marketing of video: analog electronic music. It offered two important models for rethinking the architecture of video signal processing: modularity and voltage control. Applying modularity to video on the simplest level meant that a keyer, or mixer, or switcher could be distinct machines, each with its own inputs and outputs. Video systems like the Sandin Image Processor broke down functions further into modules, such as the comparator or adder/multiplier. As separate modules, the order in which you placed these devices to create a signal path determined how images were layered, and greatly increased the number of visual configurations.

But the new designs went beyond this – they found common elements among processes that were previously considered disparate techniques. When the concepts of voltage control and analog-generated signals were borrowed from electronic music and applied to the unique time-based characteristics of video, larger patterns emerged. Keying was, in fact, another type of switching, and wipes were, in turn, another variation of keying. Within this framework, collage and montage were more of a continuum. People may have understood that the video frame was made up of two fields, and the field was divided into lines, but it was largely an abstract concept. Now, through these tools, that concept was experiential. You could apply that concept to your understanding of the material and the ability to make works.

So, the open-ended architecture has two considerations. The first invites a strong educational component to the experience of working, which could potentially be carried over to the experience of those viewing the work. It is about articulating processes rather than mystifying the viewer with effects. The second allows more choices and the ability to decide if and how those choices can manifest themselves as signature gestures of the artist working. Within this larger system, when does a particular subsystem or 'patch' become a personal expression?

Finally all of these qualities begin to merge. Being able to understand the architecture, preconceive a patch, and then perform with it in real time while allowing for elements of chance is all part of the creative dynamic. But there's more.

Indeterminacy

Indeterminacy involves the ability to put something into the system which incorporates a strong element of chance in the outcome. Indeterminacy is not randomness. Many parameters are fixed and some are deliberately left open.

Although indeterminacy is often associated with real time, particularly in the context of music as in the works of John Cage, it has counterparts in all experimental media practices. When you scratch, draw or paint across the frames of a film, use certain approaches to hand processing, or take multiple exposures by back-winding film in the camera, you are combining elements of control and chance that are unique to the photochemical and mechanical properties of filmmaking. Tony Conrad did a series in which he cooked film. William Burroughs used his cut-up techniques in writing for combinatory approaches to text. It is the pairing of indeterminacy and real time in electronic systems that offers a unique dimension for combining and modulating images and sounds.

There are many permutations of chance techniques in the ETC system, and they can quickly multiply in complexity depending on the number of devices that you add to your patch. On the simplest level, you can combine more than one tape or Quicktime file for basic collage and montage. Source images may be starting or looping at various points in time. The mix between these sources may add up to completely different combinations as these distinct loops repeat in and out of phase with one another. Other devices, such as oscillators, may be determining their own cyclical patterns of varying rates and rhythms for mixing or keying between these sources. This is why five minutes of source footage can easily generate thirty minutes of recorded material from the system without repeating the same results.

This approach is a departure from having clips locked together on the tracks of a timeline, with key frames determining the precise transition point between clips. Computers can offer many elegant approaches to indeterminacy if the software is designed to take advantage of it, as with Max/MSP/Jitter. However, the emphasis in software is often on minute details of control, grounded in conventional notions of postproduction. The design is based on the assumption that you always know what you want ahead of time, and sometimes you do. But a system should provide a balance between implementing preconceived notions and allowing a sense of play that can pave the way for happy accidents.

Interactivity, sound-image synchronization and electronically-generated images

The last three characteristics are closely related. The fourth important idea is the system's ability to allow variable and simultaneous modes of control or interactivity. There is manual control – the interactivity between person and machine, and there is voltage control – the potential interactivity among machines. The former may include knobs, switches, sliders or body movements in front of live cameras. The latter can take many forms. Paralleling their use in electronic music, oscillators allow for cyclical changes in the parameters of an image. Those parameters could include the brightness, contrast or color of an image, or the transition between two images. Those cycles of change could be happening over the course of several minutes, or at flicker rates, or faster than each field of video. Several events, or changes in parameters, could be synchronous or asynchronous. They could be completely automated while you are in front of the camera, or they can be continuously readjusted by you at the same time. These parameters could also have a very wide range of modulation, unlike controls you customarily use on a video monitor. For example, the brightness control could swing from a completely black screen through the 'normal' video image to a fully white screen. The chroma phase, or hue control, may shift the colors of an image completely around the color wheel.

The fifth characteristic of the system is its ability to establish synchronous relationships between sound and image. The desire to do this has long had roots in the concept of synaesthesia. Synaesthesia is a neurologically based phenomenon, but there is a history of trying to realize this concept in media. It includes color organs, theories of paintings by Kandinsky and Delaunay, and various approaches of early-twentieth-century European and post-war American filmmakers. With video, this was arguably the first time that sound and image had been reduced, or transduced, to the same medium – analog electronics – and that this relationship could be established in real time, where the artist could potentially interact with her own voice or image.

Before I expand upon the ways this can happen, I'd like to mention the last important idea of this system, which is the ability for images to be camera-less, or generated from within the raw material of the medium. This isn't a new idea in itself and parallels techniques in other media. In photography, there are photograms. In film, there are numerous techniques that encompass drawing and painting on clear leader or scratching on emulsion. In computers there is a whole genre of generative software. Camera-less images in video are probably most influenced by the parallel practice in audio synthesis – the ability to generate sound internally, without a microphone or recording device. All of these techniques share a common concept: that the physical properties of these media have a world unto themselves separate from their function as representational or reproductive instruments. In all camera systems, there is the question of what happens between the camera and the display, or the event represented and the representation of the event. In some ways, what we refer to as 'processing' is the questioning of that process, or a search for ways to interrupt that continuum from camera to display. The logical extension of that question is to explore what happens when we take the camera, or real event, out of the equation.

In some ways, this brings us back full circle to the first trait of the system – its real-time capability. While experimental film has dealt with these same issues of process, the ability to simultaneously and in close proximity see the event represented by the camera and its representation adds its own unique dimension. Likewise, modes of interactivity, sound-image relationship and camera-less imagery all merge with the placement of our body in a real-time environment.

Sound and image can be combined in the following ways:

- Images and sounds can be generated by the same electronic source. You can send the same oscillator (or combination of oscillators) that you have sent to the video system to see, to the audio system to hear.
- Parameters of a sound can control parameters of an image. The sound can originate from a live microphone or from a prerecorded or electronically generated source. In any of these cases, changes in the volume of the sound can be used to modulate the image.
- Parameters of an image can be used to control parameters of a sound. One way is for one or more light sensors to be attached to the front of a monitor to detect changes of brightness at specific points on a video image.
- Whether electronically generated or not, the parameters of both sound and image can be modulated by a third source, such as an oscillator, resulting in synchronized modulations of sound and image.

In any of these applications, the viewer of prerecorded work may experience the synchronization of sound and image without knowing the specific causes and effects involved. In the live image-making environment of much installation and performance, those causal relationships are much clearer to those present, either because they know the patch or they are interacting with their bodies. Sometimes it's a challenge to articulate that relationship to an audience watching and hearing the work after the fact.

Materiality/conclusion

The notion of materiality is an elusive thing in electronic media. With film, issues of physicality are obvious. It is very tangible and visible. You can hold a strip of film to your eye, punch holes in it and mark it, and see it move through a projector. You know why an image is displayed and how that image was formed by mechanical and chemical processes. You must know how and why processes happen in order to make an image at all. You cannot make film and not know something about the process. In video you can. You can shoot and edit with no idea about the process involved, because cameras and post-production software are designed to mask that very process. The result is a tendency for people not to think about the materiality of video at all. In many ways, we have to communicate with and rely on creative engineers who make machines that can be placed between the camera and the display and alter that process, so that we can develop an understanding of video's raw material.

By the time we move on to software, the processes become even more abstract. We are pushing around icons that represent our images. With most software, we are working with these highly developed metaphors based on other systems (painting, music composition, printmaking, animation, postproduction), and they inform us how to interact with the machine. Physicality becomes irrelevant because we are completely in the realm of emulation. While these programs are powerful and useful, you have to come to terms with what is gained and lost by moving from one medium to another.

An important component to this experience is the concept of signal in video. While this concept may seem initially abstract to us, our perception of it is aided in two ways: by our ability to see changes made to the signal in real time, and by the use of flow charts to document their configurations.

The ability to turn a knob, utter a sound, or move in front of a camera and see results instantaneously is the closest thing we have to tactility in analog video. It is still very different from the physicality of film, but more discernible, less remote than how we experience most computer software. This is where all of the previous elements (interactivity, indeterminacy, architecture, image-sound, etc.) come into play to help us develop a sense of materiality in electronics.

Signals can have a great deal of flexibility in how they are configured with a few basic guidelines: 1. Each signal has an origin, and in the case of video, it can either be generated from an electronic device, such as an oscillator, or converted to electrical energy from another source, such as a camera. 2. Signals have direction, and can travel in a single serial path, or split off into multiple paths. 3. Multiple signals from different sources can converge at a single point. In the case of video, this convergence point could be a device, such as a mixer or keyer. 4. Signals have a destination, and in video, these are usually the display, recording and capturing devices.

These variations determine how images will be layered, blended, sequenced, displayed and documented. The configurations of signals are called patches. Flow charts are graphic representations of patches. They are the blueprints for understanding the logic of collage and montage in analog video. Tools such as the matrix switcher and the patch bay are used to construct those patches, and are called routing systems. These tools don't process the video signals as such, but guide those signals to the video processors, such as keyers, sequencers and colorizers. Their designs allow you to split and converge signals in countless variations. Developing a sense of syntax with this system requires the skill of translating what we see on the screen into the way we read and write the flow chart of a patch.

Our understanding of how analog video works is not usually empirical in the way our knowledge of film can be. We can talk about the relationship of voltage to light, the linearity of video information, the direction of the signal, or how the two-dimensional raster is formed in the cathode ray tube; as raw material goes, video still isn't tactile, and the process isn't visible. The ETC studio gives you an excellent opportunity to experience that process and consider materiality in live electronic media.

On Voltage Control: Interview With Hank Rudolph

Kathy High and Mona Jimenez

Hank Rudolph served as the Arts Coordinator at the Experimental TV Center from 1984 until 2012, providing individualized instruction to artists participating in the Residency program. He is also a media artist creatively using both analog and digital systems to produce his own work.

Mona Jimenez: Could you explain voltage control as a concept that was and is used in the Experimental Television Center (ETC)?

Hank Rudolph: Voltage control was borrowed from audio synthesis and electronic music – as opposed to manual control. Instead of turning a knob (a continuous parameter) or flipping a switch (on/off), you could do this with an electronic signal. And that electronic signal, for something that changes continuously, could be the brightness or contrast of the signal. One source of voltage control is an oscillator, which is cyclical in nature. It's going up and down, and up and down. You could have different oscillators that are not synchronized to one another, changing different parameters in and out of phase, to create a more complex rhythmic situation. Another source of voltage control is preexisting audio – a tape or a live microphone or something, that goes through a device called an 'envelope follower', and that basically translates changes in volume of the sound to changes in voltage. So as I speak louder or softer, it can change those parameters, like brightness and contrast or whatever, the same as an oscillator can, but it's synced to the sound. But only with the parameter of volume; it doesn't change by tone or pitch. There is something called 'frequency voltage converters', but unless the sound is relatively simple, it gets confused pretty easily. Certain sounds work better than others.

MJ: How does that relate to the plus and minus system?

HR: Our system is referred to as the plus/minus-five-volt system, which is a cryptic name, but it's kind of a generic term for this system that's not quite audio, it's not quite video, but it can be applied to all of those things. I think ETC came up with this voltage range because it was compatible with preexisting audio synthesis devices. I know others use zero to five or things like that. The Sandin [Image Processor] uses plus/minus a half volt. That's closer to what video is, because video is around one volt peak-to-peak.

The Sandin really is a different kind of architecture, unlike the Jones modules. We use the matrix switcher, but it's all regular video, two volts – you know, two volts unterminated, one volt terminated – going through the video devices; and then all the control voltage inputs are mini-jacks on the front panels that take plus/minus five. So they're two completely different standards. And there are some cases where you can take a plus/minus five signal and attenuate it to two volts, so that it can go directly into the matrix as an image source; but for the most part, David [Jones] devices were designed to take the control voltages in through the mini-jack. And that's very different from the Sandin, where everything is in the same kind of plus/minus-half-a-volt soup. And there might not even be specific things on there that say 'control voltage'. If you want to pulse an image, you might take an oscillator and a video image and put them through a mixer. And then how much of one or the other is on there will determine how it pulses. So it's a little different that way.

I went to school here [Cinema Department, Binghamton University] in the mid to late seventies, when the Center was in Binghamton. I still managed to volunteer [at the Center] in exchange for using the equipment, but I think Neil Zusman and Isaac Jackson and I were among the last students to do that. Rich Brewster was building yellow boxes [for voltage control] – I was a student – and he was also building some video modules. Initially, the Center had these big army-surplus oscillators. They were these giant silver things that put out square waves and stuff. And there was this modified Sony special effects generator – I think it was modified by David – so it had other features like keyers, and then there was a Jones keyer. So I just kind of came in, would put that together with some black-and-white cameras and just use it as such. I'm trying to remember, when I first came here, what had voltage control. For some devices, voltage control was an option, but some devices don't do anything without voltage control. The Wobbulator is an example of a device that has to have voltage control. And I think when I came here, the Wobbulator did have those big army-surplus oscillators plugged into it. So the first voltage control devices were the [Rich Brewster] yellow boxes; that I can remember. (See Color Plate 41)

At the time I was trying to understand it, there was a course offered in the music department at Binghamton, with a big Moog Synthesizer. So I took that course and was learning voltage control there. There was this great essay on voltage-control synthesis by Joel Chadabe (1975). I really understood from that what voltage control did and how there are second levels of control voltages and things like that, and how to read a flow chart and, you know, all that good stuff. So then from there, I was able to apply it to ideas in video, once those modules were finished at the Center.

MJ: It does seem like those ideas were floating around. People will talk about, 'Yeah, I was a teenager and tried to make my TV image be weird by applying some voltages to it.' Or, you know, of course, Nam June Paik, with the magnets. What was your perception or experience of those ideas floating around?

HR: Well, it definitely had its foundations in music, in audio. Peer [Bode]'s father [Harald Bode] wrote a pivotal essay about modularity in voltage control in the early sixties. It's interesting that – you know, jumping ahead to Max/MSP and Jitter, and going back to voltage control – that in both cases, these started off as audio systems. This very elaborate architecture was devised for working with audio, and that the video aspect was an afterthought. The system was already in play, and then visuals, moving images, were kind of plugged into that concept. I was very interested in the idea of connections between audio and video. I remember at that point there wasn't that much to read; there wasn't that much written information on audio synthesis, and there was *nothing* on video synthesis. I was in Wisconsin in '73 and David Bordwell, a film theorist who taught there, told me about this guy, Nam June Paik, who was coming, and he uses a thing called 'video synthesizers'. That was the first time I'd ever heard of it. I saw him speak and I was just really fascinated by it. That was one of the reasons I eventually wound up at school in Binghamton.

MJ: You mentioned the Serge modules. Is that what the yellow boxes were built on? You also mentioned *Electronotes*…

HR: *Electronotes* was a newsletter published by Bernie Hutchins, who I think was an engineering student and professor at Cornell [University] in Ithaca [NY].[1] This was very, very DIY stuff. You could make your own oscillators and voltage-control amplifiers and stuff like that. It was at the same time that Bob Moog was in Trumansburg [NY].

MJ: What were the Serge modules?

HR: They were very modular audio synthesizers. There were different schools of thought about audio synthesis, and Bob Moog's machines were by far the most popular. I've read stuff about how he started out, and he didn't have any preconceived ideas of whether or not there'd be a keyboard as a control. I *think* what happened was that [users] would request certain things. And then slowly, it began to evolve into something more along the lines of conventional music. There was keyboard control. Sometimes oscillators, three oscillators that generated the sound had a common control voltage from a keyboard, so that was like a chord. Sometimes the sequencers, the devices that would sequence notes, could pick up in the middle of a sequence, so that was like a fugue. It seemed like the Moog Synthesizer, even with its great sound, was gravitating more and more towards conventional music ideas. The Serge was a *lot* more modular. In fact, with some of the devices on the yellow boxes, in order to get it to oscillate, you have to do a little feedback loop, where you have to take the pulse output of it and feed it back into the input just to get it to oscillate. So it was really modular.

MJ: Those yellow boxes were really the first boxes that were in [the ETC studio] that were built by Rich Brewster.

HR: They were used for control voltages for video, and they were also used to generate audio. But they were all, I think, based on other designs in audio synthesis. And what that also meant is that the oscillators only have a certain frequency range. When you use oscillators as a video source [in the ETC studio], when you plug an oscillator directly into that place on the matrix that cuts it down to two volts, and you can see it right through the output amp as a raw signal, when it's low voltage, it's just going from black to white. If it's a square wave, it's just black for a while, then it's white. If it's a sine wave, it's gradually rising up and gradually coming down. And there's that threshold point called '60 cycles'. That's where the oscillation, that's where the frequency, is happening faster than each field of video. So then what happens is it starts to break the image up within the field, so that so many scan lines are white; if it's a sine wave, so many scan lines are white, so many scan lines are middle gray, and so many are black. And then when you go from there, you start getting more and more horizontal lines. Then the next threshold point is when the frequency happens faster than each line of video. So whatever that frequency is, like 525 times 30. That's up around 16 kHz. And that's where it gets very chaotic and it wants to oscillate within each line of video. And that's usually where oscillators have to have a sync input. You can give it horizontal sync in order to stabilize it, in order to see these vertical divisions which occur when the oscillation happens at the same time on each line.

Those 16 kHz frequencies are already starting to get out of the audio frequency range. I mean, the oscillators that were built for musical purposes, including the Doepfer modules, weren't really designed to do video. I think the next generation of oscillators that David came up with when he built the bank of oscillators had an external sync input; and the oscillators were high enough to divide up so you start getting vertical shapes. It's the mixing of the vertical shapes and the horizontal shapes that you start making more complex shapes in oscillators.

MJ: And then from there, people would use a keyer to assign certain video to certain parts of the tonal range in the image.

HR: Yeah. It can be sent to the keyers or the colorizer or any of those things. People were already doing that; for example, Steve Beck. The Sandin had its own oscillators. And those were already up in the high frequency, where they could start to divide up the stuff. At the Center, two of the oscillators that are on the yellow boxes have a three-position sync switch that's either free-floating or horizontal sync or vertical sync.

MJ: I think the keyer and the colorizer are probably more easily understood by people who are less familiar with the [ETC] system. It refers back to regular old video language. But how do you describe the sequencer when people come to work, simply?

HR: I start off by saying it's kind of a voltage controllable switcher; and if they don't know what voltage control means yet, I say it's kind of an automatic switcher. But that's a really simple explanation, because switching can happen below and above the field rate. So you can actually start switching within the image. Especially with the binary control, once you start putting in an oscillator you can start to build up a split screen of various images. So a sequencer does a lot more than just switch. And that needs voltage control. You can manually switch, but that's not particularly interesting. With the keyer and the colorizer, I think voltage control adds a lot to it, but it's not necessary to use them; whereas on the sequencer and the Wobbulator, it is necessary to have some sort of voltage control.

MJ: And people are making more complex shapes with the sequencer – with a combination of devices, really, with the sequencer and the keyer.

HR: Yeah. A lot of these devices – the keyer and the sequencer – are really simple. But because of the way it's set up at the Center, with the matrix switcher, and the fact that the system is very modular and open-ended, it doesn't presuppose any order things have to go in. You're really starting off with simple machines and building more complex ones. So that idea of modularity combined with voltage control are really both borrowed from audio synthesis.

References

Chadabe, Joel (1975), 'The Voltage-Controlled Synthesizer', in Jon Appleton and Ronald Perera (eds), *The Development and Practice of Electronic Music*, Englewood Cliffs, NJ: Prentice Hall.

Notes

1. See http://electronotes.netfirms.com/. Accessed July 19, 2012.

"Insofar as the rose can remember..."

Carolyn Tennant

I had forgotten about the mayflies. But when I saw them outside the window, I quickly remembered the bats. And the stories of another artist who would hole up in a tent he set up in the studio during his residencies.

Barefoot, I walked towards the windows across the old wood floor strewn with cables, navigating my way through an arrangement of television monitors on rolling carts, light stands, and props used earlier in the day. A pink tinsel Christmas tree on loan from the Hand of Man, the holiday gift shop two floors below, was selected from the nutcrackers and nesting dolls spilling out of the upstairs stockroom.

It was June and we were in the attic of a riverfront store overlooking the Susquehanna, so they were no surprise. A swarm of these ephemeral insects might be taken as a positive sign, their presence considered an indicator of well-being in aquatic environments. But it was the angle and that perspective of looking into, rather than up at, the street light that captivated me and kept me at the window. Watching the frenzied mass it was hard to imagine, within such apparent chaos, how mid-flight reproduction could occur.

Down below, street vendors made last-minute preparations for the annual 'Strawberry Festival'. From the same spot, my husband J.T. and I would watch the parade the next day, as an endless procession of sports cars rounded the corner and continued their crawl down Front Street. And while we eventually made it outside, later that night we chose to watch the fireworks from the windows at the opposite end of the studio. Sitting at the edge of the bed we would take in the show's climax, a pyrotechnic strawberry that illuminated the bridge and the water below.

Among the last artists to visit the studio, we were, in a sense, present at the grand finale of a decades-old artist residency program. The significance of the opportunity and what we were there to witness was not lost on us. But to honor a space like this – equal parts electronic laboratory and artist retreat – required discovering our own methods of inquiry. We could do this by enacting its many functions for ourselves, not by attempting to retrace the ubiquitous signal that, like the river below, has run through this place.

I remember looking back at the system from the windows that night. Perhaps I was channeling the energy of those desperate mayflies, but once more I found myself consciously suppressing my desire to document every inch of the studio. I knew it was a project too Herculean to realize, and that it was compromised by an impulse to summarize something in the process of transformation.

Walking along the rough wood floors, I realized that neither mysticism nor mathematics could suggest that moment. Besides scraping a microphone along a wall, how does one sonically represent the texture of a brick? If we could map the space through abstraction, visualizing the electric hum through a palette built from the accumulated sounds of each machine, would we? Our commitment to participate in the moment – not in spite of, but because of its meaning – has thankfully relegated this fallacy to conceptual fantasy.

Insofar as the rose can remember, it has never seen a gardener die...

Like the mayflies that serve as a barometer of freshwater ecosystems, there are methodologies used to measure the effectiveness of organizations and to monitor their evolution. Matrices based on cost ratios and overhead are employed to determine the feasibility and sustainability of outcomes, but never to assess the needs that defy measurement.

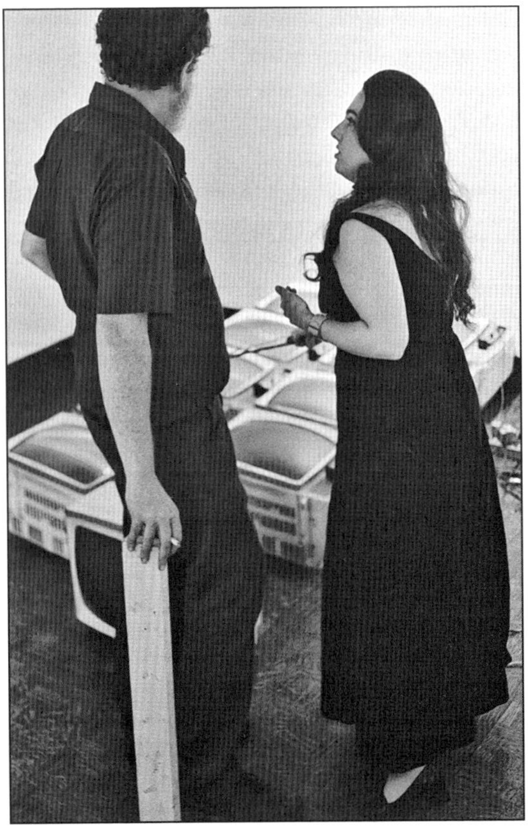

Figure 1. Ralph Hocking and Charlotte Moorman with the *TV Bed* at the Everson Museum of Art, Syracuse (1972). (courtesy. Estate of Evangelos Dousmanis).

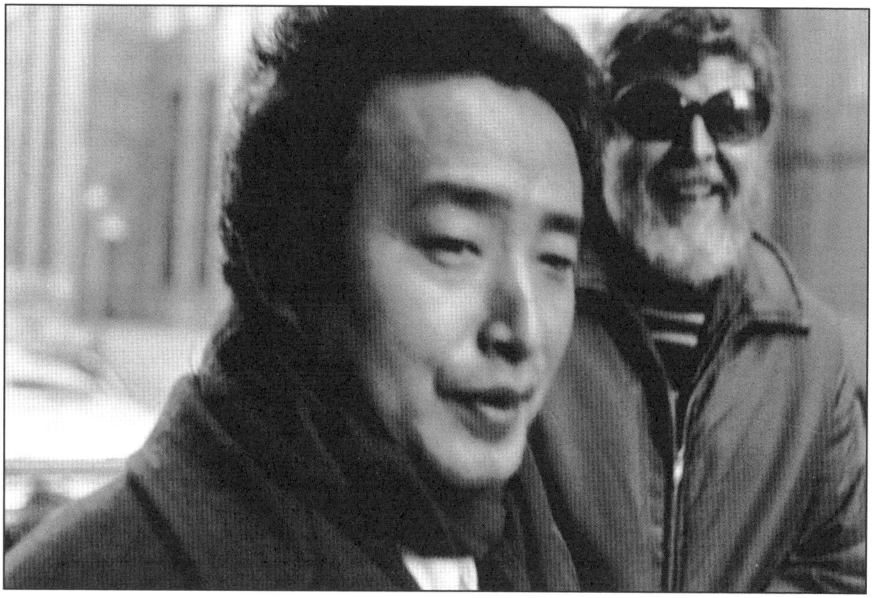

Figure 2. Nam June Paik and Ralph Hocking, NYC (1971).
(courtesy. Estate of Evangelos Dousmanis).

When the Center began, it addressed a very clear need among those artists it served: access. At first it was access to the means of production. This was not granted by duplicating the keys to a broadcast studio, but by building new spaces, and offering artists an all-access pass to the tinkerer's workshop. It soon evolved into a home for customized technology and, in situ, a system developed over time from newly fabricated tools and repurposed machines. The Center provided a space for experimental pedagogy and experiential education. For some it was a place to work collectively, so the Center provided a temporary gathering place. For those who came to work alone, the space could accommodate intense production and solitude. Whether to create art or to experiment with new methods, successes and failures were valued equally at the Center.

Over the years many of the once-limited technologies came to be emulated through software programs. Still, the Center offered resources that for artists and cultural producers remain as scarce as ever: space, time, and a safe place where art is considered work and where experimentation, with all of its risks, fosters artistic progress. Meeting these needs was an original impulse that inspired the Center's residency program, and the resources it offered nurtured artists in a way that is hard to quantify. But though the needs have not diminished, the larger environment has changed. The systems once put in place to foster this type of cultural production are, today, less hospitable. But from where I stand in the studio, on this night, I attempt to stay present and focused, aware of my perspective and the gift that parallax brings.

Back with J.T. at the matrix, we dial in our newest portrait. Though most of our experiments have not dwelled on the circumstances, one project has offered us a way to visualize our feelings and honor the spirits who have created this space. Using the pictures of those who have played major roles at the Center – Ralph Hocking and Sherry Miller Hocking, Dave Jones and Hank Rudolph – we interweave layers, mixing in Conway's Game of Life. We unleash the simple cellular automaton to render the portraits, once recognizable representations, a degraded trace of what they were.

Using these images, J.T. reminds me, is a metaphor for the system itself. Like the work produced at the Center, once abstracted from its source, the work remains rich with meaning for the artist. But the initial and final states of the algorithm have no discernible relationship to each other. It is the process in-between where meaning is either gained or lost.

An image of Nam June Paik's death mask transforms into a constellation of flickering pixels. As we turn our attention to the Wobbulator we find that Charlotte, our spider who has cast her web across the modified television, is busy with a mayfly. With only minutes left in its short life, it was no doubt pulled into her snare by the flickering monitor. Replacing Paik's death mask with an image of his late collaborator and muse, Charlotte Moorman, we turn off the Game of Life and allow the hacked electronics to animate her image. Turning on the camera to capture the action on-screen and off, we watch as they spin and twist, both Charlottes at work.

Analog to Digital: Artists Using Technology

Yvonne Spielmann

With the formation of electronic media, namely television and video, a focus on the ways in which artists use technology reveals close interrelationships and shared creativity between artists and engineers. Technical developments from the 1960s to 1980s play an equal part with artistic-aesthetic interests in shaping the electronic medium of video and experimental television. Video is considered to be a 'live' medium because of the possibility of simultaneously recording and displaying the video information; for example, the artist can control the 'image' on a monitor at the same time as the signal is being recorded. The immediate and simultaneous realization of televisual and video sounds and images in different media devices – such as the camera and monitor – also separates video from film and photography, where the image has to be developed photochemically, and can only be seen with the passage of time. However, of greater importance to the experimental endeavors by artists and engineers are the modular structure and processual form of the novel medium of video – a medium that, differently from film, has no particular order to its apparatus; in other words, there is no fixed set or sequence of tools for the generation, transmission and display of electronic information.

In the usual television experience, the viewer receives images and sounds from camera- and sound-recording processes that are structured in a particular way. Television uses a formalized set of recording and transmission standards that predefine the format of the image according to the world zones: NTSC in the United States and Japan; PAL in Europe; SECAM in France. 'Transmission' can refer to transmission either through broadcast or through communication among devices like cameras and monitors. (Video conforming to the standards is referred to below as the standard televisual format.) Analog recording technologies, in that they represent continuously variable states that are analoguous to the depiction of physical processes that were in front of the camera, technically connect to film and photography. However, considering its standard characteristics, video's position in the media system should be described as bridging between analog recording technologies on the one hand and, on the other hand, digital processuality. Conceptually, electronic signal processes foreshadow digital processuality. Also, as the development of electronic and media tools by artists and engineers progressed, it was possible to build devices with computer components and digital functions. A number of these devices were primarily designed to give the user control to drive and reposition the flow of electronic signals that had been manipulated, altered and/or processed.

In general, artists and engineers who have investigated and constructed media instruments suitable for video processing have successfully demonstrated how conceptual and technical developments intersect. The common interest of artists and engineers lies in the exploration of an electronic vocabulary insofar as they create tools that perform variations to the standardized televisual format (PAL, NTSC or SECAM), and interfere with and/or modify the appearance of the televisual medium – recording, transmission and display of audiovisual information. The exploration involves the borrowing and blending of devices or modules that can be plugged together, and also by the application of concepts from other electronic processes, such as from analog synthesizers initially designed for the audio arts. The development of an electronic vocabulary also means the expansion of the electronic modes of expressing sounds and images to the limits of what can be heard and seen.

As an electronic medium, video shares properties with broadcast television both in terms of technical specifications and with regard to their shared apparatuses. Both the signal transmission and the information compiled in the image connote a fundamental instability of the audiovisual medium. The technology of the signal transmission – whether used for video or television – generates an image that comprises a flow, an uninterrupted stream of voltages or picture elements. The video signal transmitted by the camera comprises these constantly moving elements in lines across the surface presence, or raster; the synchronization of the lines creates the format of the screen. The raster thus expresses the flow of the electrical pulses. The process of the lines being 'written' is also referred to as scanning.

In addition, two interlocked half-images constitute an 'image' in video or television; all of the information in an image is in the even and odd lines – the half-images – that intersect. A 'frame' in video is the result of two image fields intersecting with each other. The standards for each field of an NTSC image call for 262.5 scan lines (usually 256 lines are visible) for each of the odd and even parts of the image. PAL video consists of 312.5 lines for each field of video. The technical need to scan two image fields inside a frame (odd numbers from above to below and then even numbers from above to below) is to prevent 'flicker'. Through the two interlocking sets of lines, the image on the display-surface of a screen appears constant.

One aspect of radical experimentation leads to the dismantling of representational video and television toward abstraction and video noise. For example, the signal stream as a horizontal line may drift out of sync and cause deflation and deformation in the visual appearance of an 'image'. Another aspect of this direction toward signal dismantling aims to control, repeat and adjust the way an external signal or voltage is applied in the course of temporal development; the use of external signals to affect video is referred to as voltage control. It is important to rescan images that have been set drifting and to lock the signal at the end of a line of video, thus reestablishing synchronization with the image format. Among the functions of custom media instruments was to stabilize and lock manipulated images so that they could be stored and displayed using machines conforming to the standard televisual format. From early on, custom-built instruments that control, repeat and readjust altered televisual/video imagery demonstrate an internal correlation between analog and digital building blocks.

Interference with signal processing and transmission by artists has happened through the design of an instrument or has been externally executed, such as by applying electromagnetic objects to television monitors. For example, artists Steina and Woody Vasulka deliberately 'unlocked' the image, setting the signal to drift horizontally. With *Calligrams* (1970), the Vasulkas demonstrate how the horizontal hold of a video input is purposely maladjusted and the image is then rescanned. Also, in this work the camera for rescanning is set up at a 90° angle to the monitor screen, which forces the electronic structure into the vertical and stresses the instability of the electronic frame. *Demagnetizer* (1965) by Nam June Paik is also noteworthy because of the interference with the internal structure of video, but from outside. In this respect it is interesting to recall Nam June Paik's pioneering experiment to apply an electromagnetic object – a circular magnet – to the television image displayed on the screen.

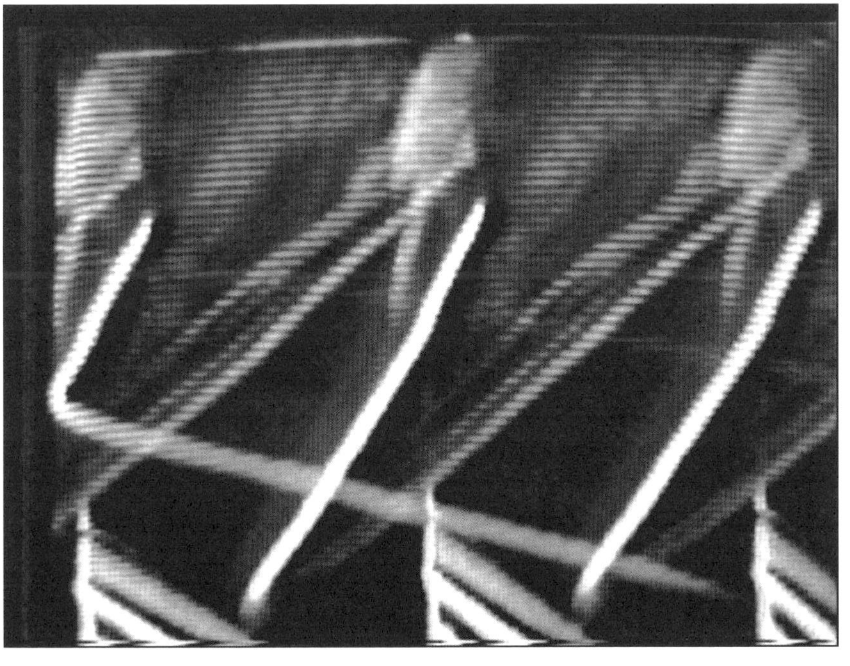

Figure 1. Still from *Calligrams* by Steina and Woody Vasulka (1970).
(courtesy. Steina and Woody Vasulka).

In the mid-1960s, the beginning era of the development of custom electronic media tools, many artists were interested in this deflection of scan lines and the violation of conventional signal transfer. With Nam June Paik's use of electromagnets applied externally to the television screen with *Demagnetizer*, he caused electromagnetic disturbances, transforming the rectangular shape of the standard broadcast image into a spinning circle. As a result, the scan lines became apparent and the two half images broke away from their usual televisual appearance as a coherent unit. In *Magnet TV* (1965), Paik also uses a strong magnet to bend the scan lines on the screen, which visually deforms the broadcasted television image so that the horizontally running electronic lines, usually invisible, are bent and twisted and become visible.

Similarly, the instrument Paik Raster Manipulation Unit (1970) is an altered television that allows the user to perform modulations of live or recorded video images, where the deflection of the course of scan lines (ordinarily left to right, top to bottom) can be reversed and become variable. Video works made on this device resemble the previous artistic manipulations with external electromagnetic distortion by Paik and others; the electronic signal appears as dancing lines. Inside this modified television set, additional magnetic fields are made possible by added deflection yokes and controls that allow the user to direct the flow of the video stream onto the screen and thus affect the shape of the scan lines. These modifications, along with the application of external audio signals, extend the possibilities of deflection, making it possible for the scan lines to be bent into various shapes, including circular shapes.

In the 1970s, Woody Vasulka was also interested in the visualization of line processing, and in the visibility of electronic waveforms in video. In *No. 25* (1975), he uses a scan processor, a video instrument that controls the width and height of an image, the intensity of bright areas of an image, and also the horizontal and vertical positions of the image on the screen. Vasulka used the Rutt/Etra Video Synthesizer (1973) to create *No. 25*. By adding an external waveform at a certain frequency, the scan processor can transform the image field, giving the illusion of a three-dimensional object; in *No. 25*, the lines were bent and stretched into a cylindrical form.

The Rutt/Etra Video Synthesizer takes an external input and, in the case of *No. 25*, the electronic information in the video image is taken from the rewinding of a videocassette. This information is treated in the scan processor, displayed on a high-resolution monitor and finally filmed by pointing a camera at the monitor. (The synthesizer has no video output. In the early days of video, some artists used a film camera and not a video camera to capture the image shot off a monitor. Vasulka thought the resolution would have been too low if recorded on video.) With the scan processor, this 'image' information is shrunk and modified to take on a cylindrical shape, which then appears as an abstract object rescanned into the 'empty' form of the electronic image (the 'video void'). The image does not convey anything beyond the presentation of the information appearing in the video void.

In addition, this video information – voltage-controlled waveforms – is at the same time displayed auditively. When the 'empty' image is curved, stretched and compressed,

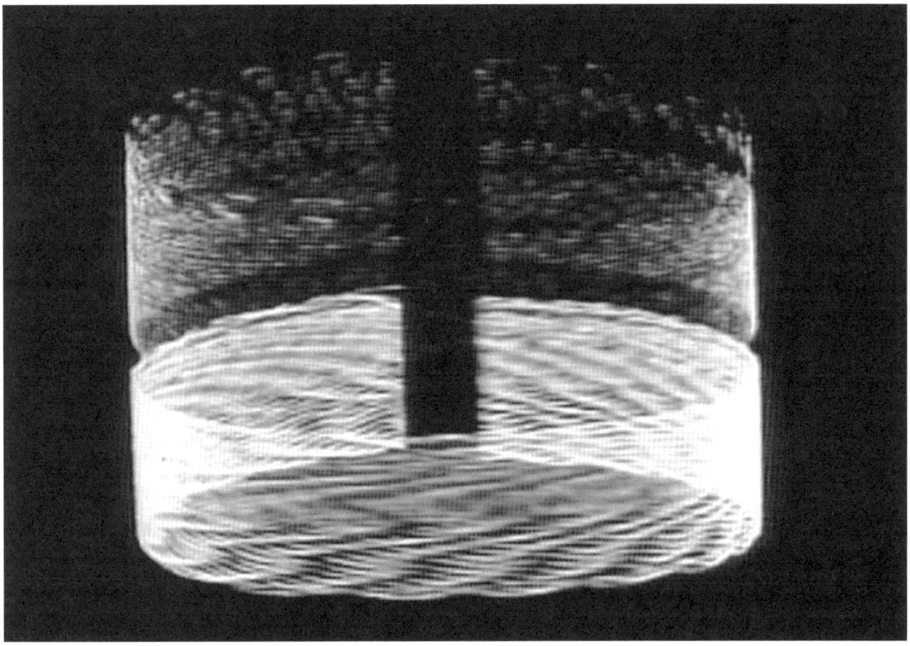

Figure 2. Still from *No.* 25 by Woody Vasulka (1975).
(courtesy. Woody Vasulka).

we see and hear at the same time 'video noise'. This video noise is grounded in the essentially audiovisual potential of the electronic medium of video. The information in the video signal can be transmitted both auditively and visually. Video is an audiovisual medium, meaning that audio signals, such as those generated by an audio synthesizer or oscillator, can be used to affect video. Vice versa, video signals can have either audio or visual outputs. More importantly, video-audio and video can be transformed into one another, and the electronic information can at the same time be heard and seen. In other words, as *No. 25* demonstrates, we hear what we see and see what we hear.

What we see in *No. 25* is a film of the scan processing of the video image and its adjustment by internal timing mechanisms, or clocks, within the scan processor. In the scan processor, one temporal operation is created by oscillators putting out audio signals at particular frequencies and voltages, a process referred to as voltage control of video. The control voltages are responsible for the cylindrical form. The effects become visible on the surface of the monitor screen when the Rutt/Etra adds synchronization pulses; in other words, the visual effects of violated signal operations occur once they are retimed/readjusted by the operation of sync that locks the violated form. Thus, the horizontal drift of the displayed lines are adjusted to the defined frame of the standard raster form. In the end, the synchronization of modified images in video to the standard broadcast signal generator is always necessary to be able to display the altered images with standard monitors. The image on the monitor electronically represents the temporal movement of the scan lines affected by processing, while the synchronization provides the timing of the flow of scan lines on the monitor. The voltage-control clock and synchronization signals are building blocks that define video's position in the media system. Here, basic programmable modules are introduced that provide the technical link between computers and video; connection exists through a clock as a pulse generator that permits the controlling of individual frames.

Another function of the instrument regards the horizontal drifting that allows the image to move horizontally through the processor. This results in stark deflation of

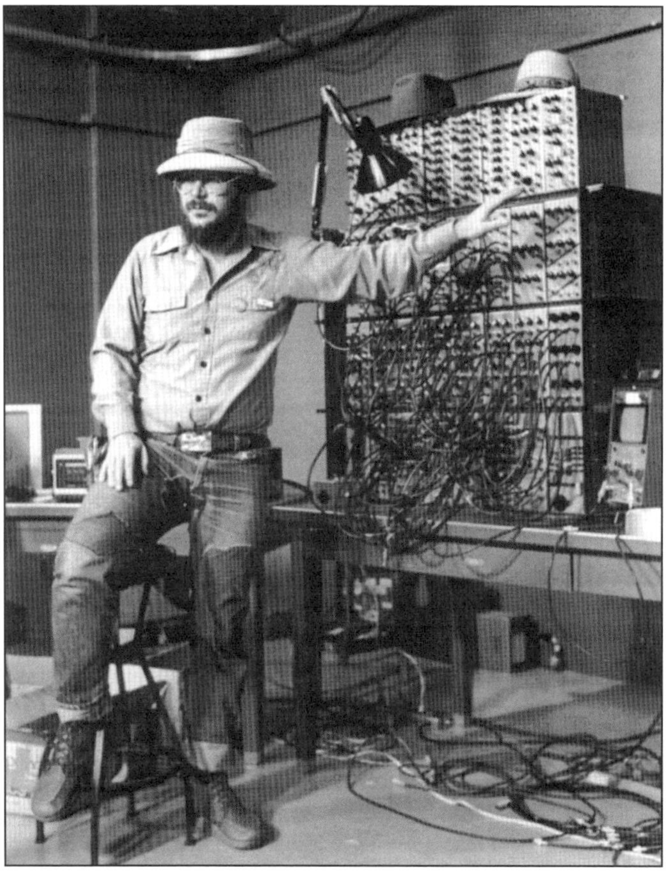

Figure 3. Dan Sandin with the Sandin Image Processor (c. 1973).
(courtesy. Dan Sandin).

the lines and causes the image to appear in the shape of waveforms. As Sherry Miller Hocking and Richard Brewster explain:

> The concept of graphic representation of waveforms is crucial to the understanding of an image-processing system [...]. The basic XY format can also be extended to incorporate a third parameter representing along the Z axis which can be conceived of as a vector extending out into space [...]. Woody Vasulka developed a technique using this type of vector diagram to locate parameters of the time frame of a video image, employing the Rutt/Etra Scan Processor.[1] This graphic representation defines the line rate, field rate and intensity information. (Hocking and Brewster 1986: 170)

Signal processing through the application of electronic voltages at particular frequencies causes changes in the shape of the raster. With one effect, when a particular control voltage is applied, individual scan lines are stretched, bent and lifted up. The brightest parts of the image are raised vertically and, as they are stretched vertically, appear wider. The contours of the image in this case mimic the graphical representation of the signal on a waveform monitor, where bright areas are higher on the scale and dark areas are lower.

By 1978, it had become possible to change the format, scale, resolution and size of the image field in individually programmed steps. With the development of the Digital Image Articulator (1978), Jeffrey Schier and Woody Vasulka constructed a machine that allowed digital processing, within which the electronic signals were sampled and constructed into discrete elements. Here, Boolean logic functions determine the interaction of signals and, through encoded processes, the look of 'images'. The digital display of the inner structure of signals differs from the display of analog signal processing and the effects of voltage control insofar as time-controlled images can now be stored and retrieved. Nevertheless, a correlation can be drawn between the analog video clock and the digital clock with regard to controlling time.

It is important to keep in mind here that, as opposed to the analog processing of signals in real time where shape and pattern of video are altered by frequency and voltage, digital processes require coding and commands in the computer through mathematical functions. Analog processors work principally with variables that are set by existing electronic voltages, and are expressed as a flow of these varying voltages over a certain time period. Real-time processes and sequences are set up through patch programming, for example with Dan Sandin's analog Image Processor [IP] (1972). Such processors or synthesizers were built to enable the user to patch modules with specific functions, thereby increasing the capacity for modifications of video signals. Although the tools are analog, they can be discussed under the rubric of machines with programmable functions. In this respect, video processing means a kind of computation, and certain optimized processors, such as the Image Processor, function as analog computers.

When Sandin describes his tool, he stresses the optional functions of computing but also makes clear that the instrument – like the scan processor – needs an external input. Such an external input is not required with a digital computer. Sandin describes the Image Processor as an analog computer as follows:

> In brief, the Image Processor (IP) is a patch-programmable general-purpose analog computer, optimized for the real-time processing of video images. [...] The IP accepts naturalistic images, modifies or combines them in complex ways and displays or stores the result. A television camera, film chain or video tape recorder or similar device can be used to encode moving images into a form which the IP accepts. A television monitor decodes the signal and displays the modified image. The instrument is programmed by routing the image through various processing modules and then out to a monitor or video tape recorder. The modules are designed to maximize the possibility of interconnection, thereby maximizing the number of possible modifications of the image. The IP is designed to accept external signals from such devices as biological and environmental sensors. (Sandin 1973)

When further discussing computation in relation to video tools, it is important on the one hand to stress the connection between patchable tools (which do offer a form of programming) and tools that are programmable through software. However, on the other hand, one must also maintain the distinction between analog and digital in order to identify the specificity and 'advantage' of digital control of the video image. What defines the Digital Image Articulator as digital is that the machine has numerical functions, and is based on binary codes and Boolean algebra. The conversion of analog to digital is thus defined: 'The procedure implies the installation of an extension card reserved for converting the video signal. Rendering in numbers can be effected continuously or image by image' (Poissant 1997: 238).[2] Here the analog signal of video is converted into discrete values that are operated on by the computer as numerical values: 'It is a case of a mode of presenting given values in the form of physical quantities (voltage, current) where the signal follows the same laws as the variations in the phenomenon studied' (Poissant 1997: 10).

It is helpful to acknowledge processing machines as analog computers 'where problem variables are represented by electronic voltages which can vary continuously within a certain range, usually -10 to +10 volts for a transistor-based machine. Electronic circuit modules allow the variables to be added, integrated (with respect to time) and multiplied by a constant' (Computer Museum 2012). The point of difference is that although analog and digital computers both operate on Boolean logic, the program can be changed dynamically with a digital computer. The results of analog computation can be shown graphically 'or be digitized for being stored or further processed by a digital computer in a hybrid system' (Computer Museum 2012). In digital computers it is the bits that define the information of a pixel, which possesses a locatable address and a definable content and color value. With the above definition of analog computers, we can, on the one hand, categorize analog tools that have programming functions and operate on the basis of plug and switch connections as computers. And we can identify digital image processing as based upon numerical methods of control, and codification and algorithmic manipulation of discrete units. At the same time, we can also distinguish between these two types of computer functions, and make their connections evident.

It should be clear by now that the borders between analog and digital computers are permeable. In a historical view, the development of computers along with other media tools needs to be contextualized within the dynamic field of media evolution that is characterized by intermedial encounters. In this respect, the new spirit of collaboration between art, science and technology that emerged with video goes hand in hand with the possibilities of modular tools – such as image processors that have patch-programmable connections and can be plugged together in a different order each time. With the introduction of the general-purpose analog Image Processor, Sandin already, in the early phase of video, managed to demonstrate a high complexity of processing video in real time, one where the patching of modules maximizes the possible combinations. And in this context of analog programming as it foregrounds the digital, the artistic exploration of both video and audio, and the experiments with their interchangeability in the electronic medium of video, just adds another dimension to the basic conditions of video as a medium of its own. The modification of electronic signals provides the medium with an optionality that is further reinforced with the optional use of instruments that can be combined and plugged together in variable order.

Shaping electronic media

Electronic culture

In the early days of experimental video, explorations into specific features of a real-time medium that could be processed live, led to the invention of multipurpose video machines that could be combined and plugged together in variable order. Shortly after the technical system of a portapak video camera and video recorder was introduced, video praxis was enhanced with a wide range of analog tools. (The date of 'birth' of the portapak as 1965 is debated.[3]) The new analog tools included audio synthesizers, video synthesizers, multi-keyers, colorizers and sequencers. In many cases, these tools were meant to be plugged together to enhance video processing, and the development of smaller and more mobile devices departed from the equipment available in TV stations.

Regarding the variety of tools, video synthesizers, analog image processors, keyers and sequencers were among the tools constructed and explored by many artists and engineers in close collaboration with each other and parallel in time to video pioneers. Particularly in the early 1970s, many of the tools were introduced in live performances to continuously shift waveforms, frequency and voltage and to alter synchronization of the flow of video signals. A common goal was to express technological determinations that were specific to video, and could express video as distinct from the languages of previously conceived media, namely television and film. And besides the possibility to continuously and dynamically change the appearance of video in analog computation, which differs from the recording of 'images' in film and photography, it was this graphical function of the processors to view the operation of amplifiers that certainly attracted video pioneers to explore analog computers for use in video. It instigated the construction of new tools but also the incorporation of digital elements into analog machines, which marked the transition from analog to digital within the medium of video and essentially foregrounded digital video.

Interestingly, the debate over analog and digital was not an issue in the 1970s when the first digital tools were available. Rather, the concern at that time was more with maintaining a position of electronic arts as art against the technical supremacy of film and photography. Both were believed to be visually of superior representational quality to the 'poor' black-and-white, rather gray and unstable video imagery. Considering the overall picture of the plethora of experiments that rework and remediate existing media settings with applications of new tools, one is struck with the observation that video pioneers were more interested in aesthetic-technical experimentation with the visual, audiovisual properties of video than in the creation of 'images' as confined through the representational tradition of painting, photography and film. Even though imbalance and tension between experimental film and elecronic arts were strongly felt (video in the first decades considered technologically imperfect compared to film), the challenging environment for video development needs to be seen in the context of the larger emerging electronic culture. The culture started off with television, a medium that, according to video pioneer and promoter of satellite television Nam June Paik, had opened a new dimension to produce 'art for 25 million people' (Paik 1984).

This direction of public media begins with the introduction of television and its leading role after the 1950s. The richness of electronic endeavors in tool development and art making also benefited from an optimistic and visionary spirit as expressed in the writings on media theory by Marshall McLuhan and Buckminster Fuller. The climate of emerging electronic culture instigated not only in-depth discussions and hands-on collaborations among artists, scientists and engineers, it also led to the groundbreaking machine performances in the 1960s of the group Experiments in Art and Technology (EAT). The theoretical reflections of EAT paved the way to an understanding of cybernetics, intermedial relationships and modular processes in a rapidly changing and interconnected media world, whereas the artists' experiments with technology envisioned the future of an electronic culture. This environment of theory and experimentation would lead to interactive and collaborative media arts and would overcome divisions of artist, artwork and spectator. The breakdown of art and media borders (through divisions of high art and low culture) also meant a more natural dealing with the quotidian use of new technologies for multiple purposes, as proposed in Buckminster Fuller's comprehensive thinking that provided a cybernetic model for hands-on production and networked machines.

So when video technology was first imported from Japan to the United States, with Sony's 1965 release of the first model CV-2000 video decks, the cultural field converging art and technology – be it in open-ended and multimedia events, performances with machines or with wearable devices like microphones and transmitters – was largely established. Influential milestone events in this direction were Jean Tinguely's staging of the self-destroying machine in the garden of the Museum of Modern Art New York in 1960 (*Homage to New York*) and performances of the EAT group led by Billy Klüver and Robert Rauschenberg. For the EAT performance series '9 Evenings: Theatre and Engineering' in 1966 (Klüver and Martin 2003), Rauschenberg had made an interactive environment from five AM radios where he asked the viewers/visitors to alter the frequencies. Rauschenberg's work offers an early example of how to use technology as both medium and tool in the participatory setting of an open art piece.

Not unlike Rauschenberg's initiative, Nam June Paik responds to McLuhan's vision of networked media communication with the videotape *Global Groove* (1973), asking the viewer to actively participate in the video process of dissection. Paik encourages the viewer to close one's eyes or half-close one's eyes when regarding the material of filtered images and sounds from television programs, programs that Paik had deformed with the Rutt/Etra Scan Processor and multiplied and modularly rearranged with the Paik/Abe Video Synthesizer. The synthesizer was designed by Nam June Paik and Shuye Abe in 1969; the one used by Paik for *Global Groove* was built at the Experimental Television Center in 1973 by Abe with Bob Diamond. Different from the standard use of synthesizers to generate the means to synthesize audio and video internally, Paik was more interested to employ his tool like a processor for the manipulation and decomposition of incoming televisual images. When viewed together, Rauschenberg's and Paik's alterations of media transmission represent different ways to perform live with the technology of mass media communication in a way aberrant from the technologies' usual institutionalized appearances. These alterations happen by defamilarizing their audiences with standardized tools, presenting radio (Rauschenberg) and television (Paik) without broadcasting.

In view of the audiovisual correspondances in video, another piece of Rauschenberg's from the same '9 Evenings' series provides an equally far-reaching concept, where the combination of audio and visual effects foreshadows audiovisual interactivity, specifically electronic interactivity. In Rauschenberg's piece, a tennis match was carried out with tennis racquets that were each equipped with a radio transmitter, microphone and antenna. When the ball hit the racquet, the vibrations were transmitted to speakers and their sounds caused the lights to dim gradually until the space was dark and the performance ended. This incorporation of radio, electronic transmitters and audio waves in the performative arts not only led the path to the early incorporation of design elements from audio synthesizers into custom video tools; it also provided an example of the intermedial nature of audiovisual technology and its crucial performative qualities.

This approach to interactivity with video was in particular highlighted in the music and machine performances by Steina Vasulka that intentionally departed electronic pictorality from the image form as an entity formal unit. In performances of *Violin Power* (1970–78), Steina used primarily the Rutt/Etra Video Synthesizer in conjunction with the Frequency Shifter designed by Harold Bode (1964), and the Multi-Level Keyer designed by George Brown (1973). The performances demonstrate how the sound of a violin performed live governs the display of the video signal: the signal displayed on the monitor screen is a deflection created by the sound of the violin simultaneously with its performance.[4] Not only are the audio and video interacting, also the artist performs the internal linear signal structure of the electronics through the tool of the processor in such a way that the performative capacity – essentially the live video medium – is forced into visible and audible appearance and expression. Experimentation with televisual behavior of images and sounds, in the strict sense of performing the specificities of a medium, necessarily drives video apart from representation toward abstraction. The tools that were built to support this direction amplify the scaling, speed and pattern of the electronic process, and cause disturbances to the standard televisual image.

Returning to Experiments in Art and Technology, the group's public events formed the path to media artists interconnecting experimental performance and innovative machines. Billy Klüver set the tone when describing the balancing act of involving industry support from Bell Telephone Laboratories for '9 Evenings', and 'modifying custom-built equipment'. As the program notes reveal, the most ambitious tool was a Theatre Electronic Environmental Modulator (TEEM), designed by Per Bjorn and Robby Robinson specifically for use with '9 Evenings'. As Klüver explains:

> It would, however, be foolish and irresponsible, to describe this equipment as terribly extraordinary in technical terms. Compared to the missiles at Cape Kennedy and the large computers it is peanuts. This is rightly so. The artist cannot be expected to make use of the most sophisticated aspects of technology, even if he [has] access to these, since he is confronted with a new material. What gives our equipment its unique value is that it was built for no other function but to be part of the performances. The equipment is built from scratch and is a result of direct interaction between artists and engineers. (Klüver 1966).

As noted above, by as early as 1973, several image processors and keyers had been built from scratch as specifically designed video tools, and as opposed to modifications of machines that were already in use. These novel machines were built in small numbers and by artists themselves, or by others in close consultation with artists who requested specific features for modifying and controlling electronic processes. One problem artists and engineers were trying to solve was to be able to duplicate, repeat or capture a processes that could only occur live. For example, some artists were interested in tools that allowed for the repositioning of the frame or that could repeat effects such as the horizontal drifting of scan lines.

One of the early tools to help control the effect of 'unlocking' the video stream from its conventional state was the Horizontal Drift Variable Clock by George Brown (1972). This device is a pulse generator, which determines the extent of the horizontal drifting and allows the user to control the deflection of the horizontal synchronization signal. Without sync the image deviates from the raster of the standard television format. Once the video 'image' is altered and set adrift, it needs to be repositioned through a rescanning process in order to be displayed in conventional video systems. As already described above, the clock, generating synchronization signals through an oscillator, is a necessary step in the development of media tools for the retiming of violated images. The clock is also necessary for the repositioning of images with, for example, the Multi-Level Keyer, when many input sources are arranged in layers according to changing priorities. As noted above, in 1973 George Brown developed the Multi-Level Keyer; this tool has a built-in clock and possesses a form of elementary programming to define a hierarchical arrangement of as many as six different layers of source video.

In summary, these unique tools – together with the adaption of existing equipment such as audio synthesizers – express the spirit of electronic art making, and served the need of some video pioneers to deliberately control the flow of electronic waveforms. These artists wanted to take advantage of modules with different functions that could be plugged together in a variable order. By the end of the 1970s, these general-purpose analog computers were enriched, and digital imaging machines were built that would emulate the analog processes but would produce video systematically, as a result of programming systems based on software. An interesting novelty that occured around 1980 was that analog-digital converters became available and could transform video information into numerical code that could be modulated regardless of its original form. By cutting off the relationship of information to its original physical manifestation, we enter the area of technical simulation, and as a result, video is completely set free from camera obscura and pinhole perspectives. Remember, this freeing of video was already an achievable goal in most of the custom electronic tools. However, simulated, nonrepresentational images, sounds and synthesis were not fully controllable through features in analog synthesizers; digital design elements would add accuracy and repetition, and allow for storage.

Artists and engineers

As noted above, video experiments of the 1960s to 1980s, particularly those that connect and incorporate both analog and digital image processes, exemplify the specific position of video in the media system as a bridge between analog and digital tools. The connection of video processes is from analog systems (patch-programmable) to digital programming systems (with the capacity to emulate the analog computer on a digital machine). This connection proves itself most relevant for the articulation of an electronic vocabulary. In this respect, video experiments utilizing such tools as synthesizers and keyers that are based on fundamental variations of waveform, synchronization and feedback contribute to the specificity of video. The vocabulary of video is further enhanced and enriched with digital tools which build upon such experiments.

In light of the attempt to articulate the specificity of the video medium, a detailed discussion makes sense concerning art making using equipment embedded in the larger media system during the period of the 1960s to 1980s. During this time span, new instruments for use with audio and video were conceived and built, which interconnect and exchange the auditive and visual energy of the electronic signal. To fully understand the audiovisual nature of the electronic media, one can look to the underlying conception of sounds as introduced by John Cage in the 1950s. For Cage, the composition 'of' sounds would be of higher relevance than the composition 'with' sounds. Prior to the 1950s, Henry Cowell had analyzed noise-sounds and in 1930, coined the term 'clusters' to characterize the phenomenon of chords of tones that merge and cause noise effects (Cowell 1969). The understanding of noisy sounds was influential in audio experiments that focused on sounds and clustering of tones rather than on music with a traditional sense of composition. What connects sounds and visuals at the structural level is the processual nature of the electronic technology that characterizes the raw material as noise. In this respect, video is just another kind of noisy sound, and experimentation in electronic music can be seen as a precursor to video experimentation. This is especially true in that synthesizers and various oscillator-based tools were used in audio arts for the modification of sounds.

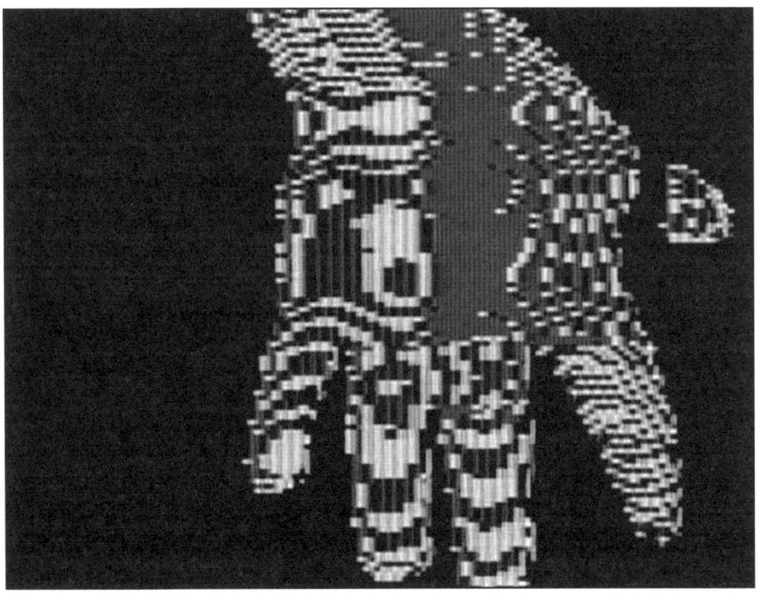

Figure 4. Still from *Artifacts* by Woody Vasulka (1980).
(courtesy. Woody Vasulka).

Regarding video's raw material of noise, as discussed above, input and output energies of audio and video can be dealt with as interchangeable impulses to process the waveforms of electronic signals, and instruments can be connected at the same time to hear and see the graphically displayed modulation of frequency and voltage, the point being that tools for manipulating the size and form of the waveforms colorize the image and/or multiply the possible modifications and number of sources that can be layered (such as with keyers). When viewed together they express an operational mode that aims to control the smallest information unit, which is the electronic (audiovisual) signal.

On a structural level, analog instruments can be compared to computer programs which control bits and digitally alter the address data (location) and value of pixels. The link between analog and digital processuality of the smallest unit in each mode can be seen in the control functions of machines of both types. It can be argued that the more analog machine operations result in control over the maneuverability of the smallest units, the more these operations can systematically interfere with the standard televisual raster format. In addition, the more effectively these operations manipulate and even violate the inner structure of the electronic medium, the more they lay the groundwork for digitally controlled effects, respectively the maneuverability for discrete processing of individual image segments.

One interesting link between analog and digital tools is the incorporation of the clock in the early keyers of the 1970s, whereby they control shifting priorities. In particular, George Brown's Horizontal Drift Variable Clock described above belongs to a category of programmable devices designed to provide an external source, or clock, for synchronization of a video signal that controls the speed and extent of horizontal drifting of the image. The device's programmable function secures the repositioning of the image after it has been set adrift, and allows the subsequent relocating and layering of image sources. Finally, a control clock is needed for keyers to solve the task of prioritizing a number of sources that are being layered. Furthermore, synchronization of the video signal through clocked pulses is necessary for any raster to appear in the standard television format. Thus, the clock function means that the synchronizing signal pulses can be demonstrated and repeated in every video display system.

To conclude, I would now like to discuss tools that are at the core of the passage from analog to digital; these are tools that permit control of real-time processes which are inherent to video. In this context, control means to systematically carry out the same set of functions and be able to repeat the same effects of those functions.

As demonstrated above, similar to any other new technical medium, video developed from technology to medium in various steps. In particular, the establishment of video's media aesthetic as an art form with an electronic language of its own paved the way for the informative intersection of analog with digital media tools. The development of this new art form resulted in the conception and construction of digital functions of devices that were not attached to, but built into, video equipment. The aim of building novel devices was to enlarge video's capacity and thereby also distinguish the manifestations of video from related technical analog media, such as film and photography, in terms of the scale of the form, the tempo of display, and the pattern of appearance. This direction of tool development naturally leads towards the convergence of analog and digital computers.

In particular, video devices, such as clocks and keyers that incorporate elements of digital control, point to the direction of processing smaller units. But where the analog clock's synchronization operates within the linear mode of the electronic waveform, digital computers' operation on bits affects pixels as the smallest digital units. What becomes evident in early video instruments is that these analog devices have programming functions through plug and switch connections. As noted above, basic programming functions were built into real-time processors such as George Brown's Multi-Level Keyer and Dan Sandin's Image Processor.

The IP's analog programmable functions become evident in Sandin's videotape *Triangle in Front of Square in Front of Circle in Front of Triangle* (1973). Sandin performs keying involving image prioritization with simple geometric forms. Through their arrangement in the Image Processor, the geometric objects display layering that appears logically incompatible. Sandin describes the video as:

> [A] demonstration of the fact that thinking of video keying as putting one thing in front of another is inaccurate and limiting. The Analog Image Processor was programmed to implement the logic equations: if triangle and square show triangle, if square and circle show square, if triangle and circle show circle. (Sandin 2004)

While Sandin presents the concept behind keying using the IP, the main creative source for his tool invention was the shared research environment of artists and engineers. Sandin, an artist and educator, started the Electronic Visualization Laboratory (EVL) at the University of Illinois at Chicago Circle Campus in 1973 with computer scientist Tom DeFanti. From long experience Sandin argues for the necessity of converging the expertise of artists and scientists as follows:

> The idea of artists bringing creativity to scientists is insulting to the engineers and the scientists, who think they are already creative. In our view, media artists should be supported by sciences because media artists share visualization technology with science; artists are trained in this technology; artists are trained in a range of visual studies; artists know about presentation, and artists are good project organizers. Finally, artists create new media, new ways of working with media. These are the reasons why artists should be supported within the research environment. (Sandin, DeFanti, Kauffman and Spielmann 2006: 220)

Another far-reaching example of collaborative research is the collaboration between video pioneer Woody Vasulka and programmer Jeffrey Schier on a fully digital device, the Digital Image Articulator. The instrument was developed during 1978–80, and the tool contributes substantially to the transition from analog modulation to digital processing. The leading concept of the collaboration was to create a tool to display the inner structure of images and sounds. Here, microprocessors were capable of real-time image processing, and the Digital Image Articulator was an approach to 'look behind the images'. The point was that the system could treat the analog-to-digital converted image information as encoded processes, and change the pattern of appearance on the matrix level of algorithmic information. Finally, the machine displays patterns of instantly moving images that allow for a wide range of different densities, and through the insertion of color and texture allow a certain number of bits to be processed in real time. The Digital Image Articulator displays the binary codes in the same interlaced fashion as analog video, with two fields of 262.5 lines per field, the standard televisual image format. In addition, one can say the tool demonstrates the distinction between switching and programming. Plug-in tools characterize analog technology, whereas digital computing is defined by encoding and by programming through all-purpose microprocessors.

To further explain this distinction, we can look more closely at Sandin's Image Processor, which uses cables and plugs to patch, and whose effects could be performed live. To stress the distinction, patching is an early or simple form of programming. Another step toward programming is taken with the Digital Image Articulator, whereby the interface to the LSI-11 microcomputer allows connection between its microprocessor and the video processor to request usage of a particular buffer. As stated in the manual for the Digital Image Articulator, digital encoding differs from analog plugging and switching not only technically but also conceptually.

In raster scan graphics, two schools of thought were prevalent when dealing with image formation. One is the processing view, where signals are seen as real-time signals that may be delayed, modified, or switched, but must conform to the restrictions of 'real time'. The other approach is the buffer or storage mode, where information is taken in and stored as sequences of still photographs and replayed or recalculated as a memory array. (Vasulka, Schier and Moxon 1979)

Once the buffer request sequence is successful, a block of data may be written into or read from the buffer. The control of the computer and the code also means that the LSI-11 is responsible for setting the buffer priority registers on the eight image buffers. In light of the technical level of digital image processing that was available at the time, Woody Vasulka has rightly expressed his view in the work *Artifacts* (1980), a series of studies to demonstrate the digital tool – a tool where one had to share the creative process with the machine.

Categories of instruments

The categorization of instruments needs to be further differentiated, as some of the tools created in the early days of experimental video have misleading names. There is a lack of precision in definitions to differentiate synthesizers from processors, which has its roots in the naming of instruments by the creators. These pioneers knew exactly what their tools were capable of, and were less interested in categorial determinations that exist in the debate around video. For a better understanding of the different functions of the tools, we need to keep in mind that an audio synthesizer is a device with which sounds can be generated; as noted above, audio sources can be used to synthesize the video signal as well. Furthermore, synthesizers (audio and video) are essentially used for compositional processes which happen inside the machine and generate a video signal at the output. In contrast, a scan processor works with external input (which is also needed in the keyer), and is designed to analyze and modulate the incoming light values. It processes electronically the information coming in from outside. Both systems, synthesizers and scan processors, have functions to cause a shifting and drifting of the signal processes in time, and can manipulate them live and continuously.

Given this definition of different kinds of instruments, predominantly the devices of Eric Siegel (Electronic-Video-Synthesizer [1970]), Stephen Beck (Direct Video Synthesizer [1970]) and Dan Sandin (Analog Image Processor) would be included in the category of synthesizers. In the synthesizer, the signal passes through various interlinked modules, all of which have input and output connections, which means that they are received and transmitted. Once we look closer at the actual instruments, their operations and the ways in which they are employed in video practices, it is necessary to further differentiate. If, by definition, a synthesizer is designed to usually generate and synthesize the signal, then we have to limit the category of synthesizer to the devices designed by Beck and Siegel. By contrast, the devices of Dan Sandin and the Paik/Abe Video Synthesizer (1969) work with external inputs and do not generate signals. From this viewpoint, it seems more appropriate and less confusing to place Sandin's Image Processor and the Paik/Abe Synthesizer – despite the name of the tool – into the category of processors. They do not synthesize a signal but depend on an external input signal, an external source of imagery like any processor.

In view of these peculiarites, the so-called Paik/Abe Video Synthesizer – which was conceived by Korean fluxus artist Nam June Paik and Japanese engineer Shuya Abe – works like a processor with video inputs and thus external image material. The ways in which Paik uses the instrument are best expressed in the WGBH studio sessions of *9/23/69 Experiment with David Atwood* (1969). (WGBH is a public television station in Boston, Massachusetts that offered residencies to artists from the late 1960s to late 1990s in the New Television Workshop.) The Paik/Abe Synthesizer produces a visual collage from manipulated electronic signals and recorded television material. The television material is cut up, deconstructed, defamilarized and compressed, with superimpositions and segments altered in intensity and color using luminance values. The result is an abstract, colored, moving texture. Generally, Paik was not interested in analyzing and manipulating signal processes themselves, for which a processor is usually applied, since he rather works with the premise of decomposition and recomposition of existing medial images taken from television programs. In describing the Paik/Abe Synthesizer, Jim Wiseman states:

> Basically, the Paik/Abe Synthesizer is a device designed to accept 14 different inputs from either black-and-white video cameras or audio tone generators. It translates these images into color video images and/or patterns which may be displayed on a color video monitor or recorded on videotape. Controls are provided for mixing up to 7 inputs at one time and for altering the character and color of the final image which can range in tone and color from exaggerated black and white through pastels, to highly saturated electronic colors found nowhere else in nature. Alterations in strength and character of any of these inputs or changes in setting of the controls will produce obvious changes in the whole output image. (Sandin 1973)

In view of novel devices that were built specifically for the performance of video, Bill Etra's description of the analog Rutt/Etra Video Synthesizer reflects a pioneer spirit in developing these custom machines, which were intended to be marketed as a standard tool, but were not successful commercially or in the mass media:

> I knew almost nothing when I started. I knew you had to sum the waveforms. That was obvious from the oscillators. I knew you had to attenuate them, which is multiplication. Steve [Rutt] knew about diodes, resistance networks etc. The first machine we built was really deflection on a regular oscilloscope. [...] I thought it was going to cost under $5000 and be sold to artists and schools. [...] The price went up because we tried to sell it to broadcast engineers who couldn't use it anyway. They didn't have the initiative to use that sort of complex equipment. (Etra and Rutt 1973: 136)

The Rutt/Etra Scan Processor and Dan Sandin's Image Processor were each only produced in less than 20 units. While Sandin – who also made the IP schmatics available for free to the general public – used the tool in his own individual video works, the Rutt/Etra was largely used by video artists. Gary Hill employs the tool in *Picture Story* (1979), recombining dissociated linguistic elements (text, image and language) and corresponding visual and reading processes by reversing the left-to-right orientation and presenting the text upside down. (See Color Plates 24 and 25.) The Vasulkas were more concerned to study the deflection of scan lines, and to present waveforms as three-dimensional configurations. The geometrical forms resulting from experiments by the Vasulkas resemble distorted screen surfaces, the outlines of which can represent, among other things, rectangles, ellipses, cubes and cylinders.

The Rutt/Etra Scan Processor manipulates the information of the camera input in real time, processually and electromagnetically, and allows the electronic signal to deviate from the usual linear sequence and the raster form of the frame. As discussed above, through one effect used by the Vasulkas, the brighter parts of the 'image' information are lifted up so that the horizontal lines deflect vertically and create sculptural forms. The deviated 'image' then needs to be controlled through synchronization within the raster frame. Bill Etra explains:

> The important contribution of the Rutt/Etra was in the concept of retiming of the image frame relative to the incoming sync. [...] The standard modules consisted of waveform generators, summing amplifier, and ramp generator. The ramp programmer was optional, allowing the user to do more than one move. The Joystick X, Y, Z controlled was also an option; it was a manual interface which could be assigned to control other devices. (Hocking 1986)

A decisive advantage of a graphic system like the Rutt/ Etra Scan Processor consists in the ability to reposition the image. This function has a spectrum of variations including the zoom, rotation of the image, and reversing of the right-left and upper-lower relationships in the image raster. The visual results of adding energy and multiplying variables in the scan processor, for example, are shown in real time on an oscilloscope which is built into the device.

Inside the Rutt/Etra, a 'ramp generator' governs the voltage in the synthesizing function, and is responsible for generating sound and image, an audiovisual simultaneity of the electronic information from the same source. This source creates at the same time

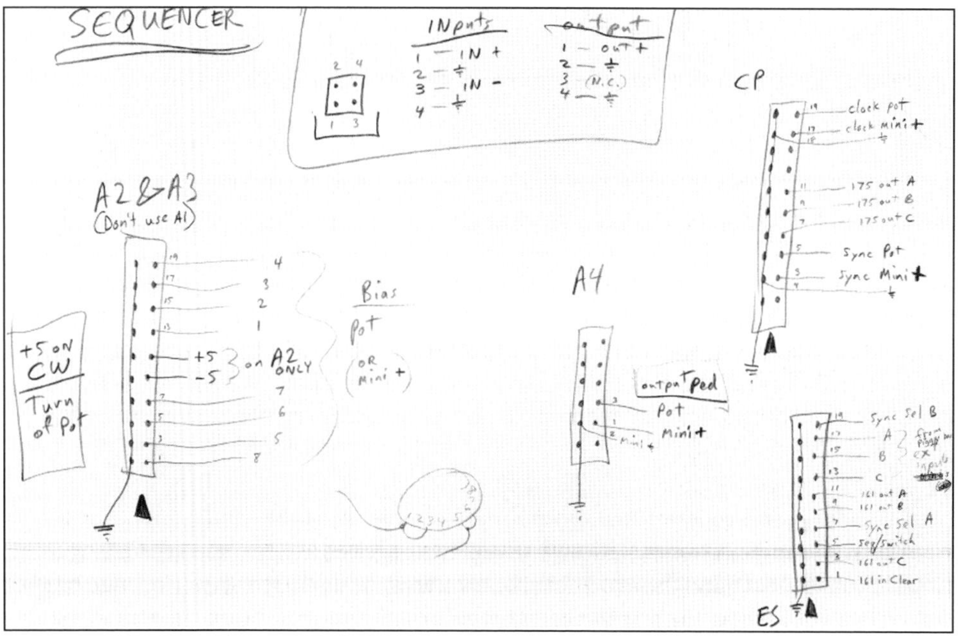

Figure 5. Drawings by Dave Jones for the Jones Sequencer (c. 1980).
(courtesy. Dave Jones).

the shape of the visual pattern and the electronic sound of this particular visual signal. An oscillator, or waveshape generator, is responsible for reshaping the pattern of the source: 'These waveforms are also used to reshape and animate external images being processed in the synthesizer. When used in combination with other waveform generators or ramp generators, it produces waveforms that are constantly moving, or ones that change from one state to another upon command' (Etra and Rutt n.d.). Thus, the same source is affecting sound and image simultaneously.

In this phase of cocreativity, artists and engineers wanted at the same time to understand and shape the specific mechanisms of the developing medium, and thereby dynamically define the range of its aesthetic language both within and through modification of the technological settings. For example, the tool designer Dave Jones was involved at Experimental Television Center in Binghamton, New York in making the connection between analog and digital processes. He participated in research to interface the LSI-11 computer into the Center's studio in an attempt to have the computer create and control video.

In addition, Jones designed the Jones Sequencer in 1975–76, which could take voltage control inputs as well as be operated manually. The Sequencer instrument can switch and select from eight separate inputs. Not unlike the above-described processors and keyers, if not being manually operated, the sequencer needs a clock (a pulse generated from an oscillator) to define the positioning of the images in time and to vary the rate of switching. As the system manual explains:

> When using voltage control, the rate of the sequence, as determined by an external clock such as an oscillator, can range from extremely slow with vertical interval switching, up to several multiples of the horizontal frequency which allows juxtaposition of signals within the divisions of the raster. (Hocking, Brewster, Bode, Rudolph and Schlanger 1980a)

Interestingly, different artists, such as Nam June Paik, Steina and Woody Vasulka, and Gary Hill in the United States; Robert Cahen in France; and Peter Donebauer in the United Kingdom had used similar or even the same type of effects machines to generate electronic pulses, synthesize images, and manipulate time – that is, control – the interlacing of different parts of the image. For these artists it was interesting to work with video because of its processual structure. In this context, tools were applied mainly to use video's brightness for alterations in the appearance of the visual form, to shift frequencies for feedback and delay effects, separate the scan lines, and unlock the signal as it runs in lines during the scanning process, such as horizontal drifting in Vasulkas' *Discs* (1971) and other segments of the videotape/installation series *Matrix* (1970–72). Like the Vasulkas and Hill using the Rutt/Etra Scan Processor in the United States, Cahen was interested in the aesthetic display of oscilloscope effects and worked with waveforms which were combined modularly. In the United Kingdom in 1975, Peter Donebauer, with his Videokalos Colour Synthesizer, conceived a device that enabled him to perform

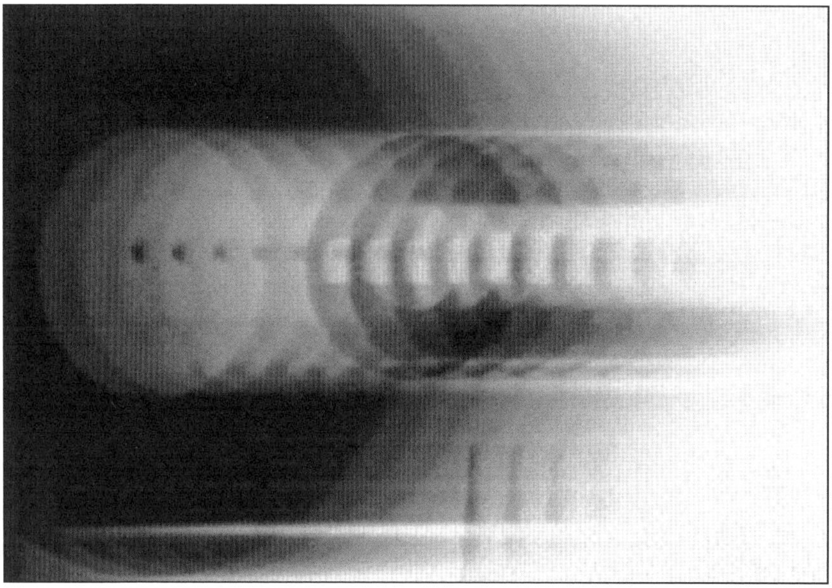

Figure 6. Still from *Discs* by Woody Vasulka (1971).
(courtesy. Woody Vasulka).

video live. In open processes, he demonstrated the synchronization of audio and visual elements towards an electronic syntax.

A striking example of image processing can be found in Woody Vasulka's study *Vocabulary* (1973), which combined the Rutt/Etra Video Synthesizer and a keyer. The work already belonged to the dimension of simulation and thus foreshadowed effects with digital machines. A keyer operates with changing luminosity values and defines the threshold between two video sources combined in the same image. With such a luminance keyer, certain bright or dark values can be removed and replaced with a second video source. As a result, new spatial relations between represented objects emerge that are based on light values, and depart from the actual positoning of the objects in space. There is a simulation of spatial appearance within the constraints of time-based analog media, made possible by tools such as keyers which provide early steps of programming and thereby bridge analog and digital.

In the broader sense of computer-driven simulation, we need to specify the relation of processing with analog electronics to the programming, adding and multiplying in the binary mode. Connections in digitial modes are made of 'or' and 'and' commands, and they also include the possibilty to reverse the processes – as in digital morphing. This option for reversal is not possible in the analog realm where the analogous relation to something in the physical world (the light information that is translated into the video signal, for example) cannot be negated. In the digital realm, it is possible to endlessly multiply and reverse applications in all directions and dimensions. In contrast, the analog process is bound to linearity, and analog video belongs to the category of time-based media. More important when comparing analog and digital media machines is the fact that, with algorithms controlling the commands for carrying out mathematical-logical processes, the artist's creativity shifts. He or she does not need to know how the program is translated into a sequence of commands which the machine performs. In this categorization, analog media tools are bound to a physical relation between the represented information (light) and the representation (the signal). Here 'analog' means in accordance with events in physical reality; the artist can interfere directly and alter, distort and almost destroy this relation, visible in the disturbing images of physical objects such as a hand and a sphere in seemingly impossible physical relations in *Vocabulary*. These kinds of relations – representing 'something' – are not relevant with the mathematical-logical expression of audiovisuality in digital computers.

Plug-in and programming

Once the 'image' had been recognized by artists as a real-time event, and defined at its most basic level by values of intensity, it could be organized according to any priority. Other kinds of instruments with specific purposes were needed as digital systems to control real-time processing, and to store and retrieve the larger amounts of information needed for real-time video than, for example, text. The Frame Buffer (1977) designed by Dave Jones, provides a necessary function in digital video processing for storing information in memory. When an image has been converted from analog to digital, and a discrete numeric value is assigned to each luminance value, the digitized information

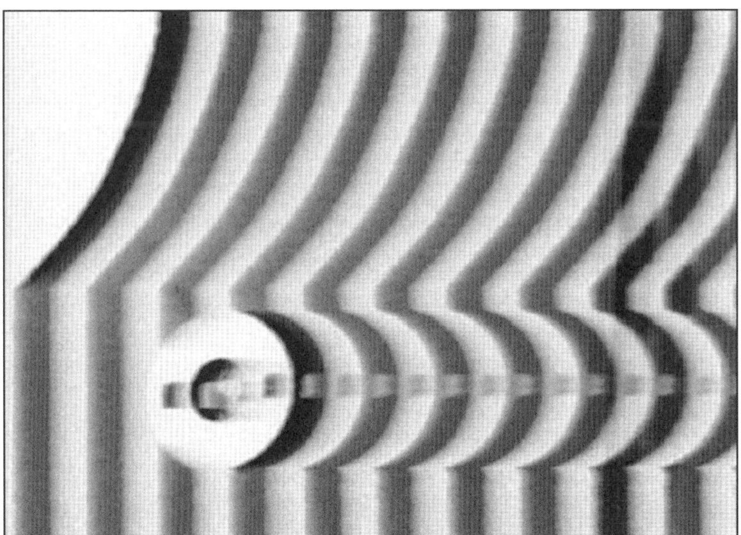

Figure 7. Still from *Discs* by Woody Vasulka (1971). (courtesy. Woody Vasulka).

needs to be stored by memory buffers. A buffer that temporarily held a complete frame of video was referred to as a 'frame buffer'. As the 'Studio System Manual' of Jones's Frame Buffer reads:

> By a process called selective storage, the buffer can grab a single whole image or selective sections of an image. Using a second video input image as a reference, a keyer-like device determines what part of the image, in terms of gray levels, will be updated with new information to be stored and what part will remain from the last storage process. Using the computer, you can also write on top of the stored information or create and paint new images. (Hocking, Brewster, Bode, Rudolph and Schlanger 1980b)

Since before desktop computers were invented there was no way to store video digitally; any digital signal needed to be converted back to analog. The signal would then be further processed, if desired, with other electronic media tools and recorded to videotape and/or displayed. The question of an image, then, that was inherited from frame-bound filmmaking and had already been challenged in electronic transformations, loses its importance. The digital task is to build an image system with its own vocuabulary. With foresight, Woody Vasulka had made the prognosis of abstract, nonrepresentational, digitally-processed video information in 1978 when he said: 'At this point it may sound almost popular-cultural, but that's the fight between reality, and the beauty of the real, and the beauty of the artificial. In some instances the beauty of the artificial has already won' (Hagen 1978: 23).

In the context of electronic culture, clearly video's position should be stated as deviant from the pictorial image, and with critical reference to any generic determination of text, image or sound. More precisely, the general positon of video culture can be formulated in light of technical simulation. Simulation is assigned to the digital where it means the potential to all images made possible through numeric processes. To differentiate the specifics of electronic and digital media forms, French media theorist Edmond Couchot (1998) reserves the attribute 'simulation' to a digital image of simulation because digital is based on computing operations. It is further assumed that when all possible forms of images are realizable in the digital, the specific electronic position of image and sound signal processing can be restructured as technical simulation. It is possible to shift an image constituted through analog processes to a numerical image by breaking its luminance values down into numbers.

On a numerical basis, then, endless variations and combinations confirm the concept of 'every image' that contains the potential infinity of all images. Gene Youngblood has made the fundamental distinction between the analog recording principle shared by electronic and cinematic registration of light pulses on the one hand and, on the other hand, the digital processing of encoded information shared by digital video processing and computer graphics. He states:

> In electronic cinema the frame is not an object but a time segment of a continuous signal. This makes possible a syntax based on transformation, not transition. Analog image processing is one vehicle of this particular art – for example, Scan Processors. But it becomes even more significant in digital image synthesis, where the image is a database. (Youngblood 1989: 28)

The concept of image synthesis was explored in synthesizers and processors: the electronic medium could be perceived as signal-based and its raw material was identified as pure energy, noise. In this respect, both electronic music of the 1950s and video art of the 1970s worked with time energy in waveforms toward abstraction.

Also, media philosopher Vilém Flusser (1985) has identitifed an essential musicalization in the electronic medium because of its tendency to abstraction. As distinct from categories of optical inscription and iconic images of film and photography, musicalization stems from the mathematical-abstract model and concerns abstract principles of organization in musical composition and technical images. The tendency to abstraction is rooted in structural relationships between image and music. When video is understood as being the signal, its inherent property is the energy that can be circulated (closed-circuit), fed back into the system (to cause feedback effects), and alternatively realized as video and as audio output. The resultant energy images are not necessarily pictorial but express abstraction. In following Flusser, it would be appropriate to abandon the model of the frame and discuss synthesized video and any operational modes of video instruments in terms of musicalization, which means working with signal processes, image or energy fields and transformation imagery.

Analog to digital

In sum, processual elements in video systems show structural analogies that connect electronic and digital tools on the level of shared modular systems and comparable processual transformations. Also, because of the interchangeability of audio and video signals and in view of video's technical option to synthesize a television signal from raw electronics, it can be argued video and computer tools have more characteristics in common than video and other analog media, namely film. There is no such thing as a 'fixed' frame (unavoidable in any photo-chemical treatment and camera-less recording of light on the film strip) and no indissoluble interval in the electronic medium – only the deliberately adjusted form of electronic information that is displayed according to agreed-upon parameters of the standarized televisual format. The video image is essentially open to drifting unless locked, respectively non-adjusted. As discussed above, discovery of video's behavior led to a series of experiments with horizontal drifting (in the Vasulkas' videotape *Discs*, in addition to those noted above), and has been formulated into the statement that video is not restricted to frame-bound images, such as film, but generically frame-unbound.

When viewed together, video's distinct features became apparent in image processors that would rearrange electronic processes, show deflection of the raw material of video noise, and express the electronic signal's potential of audio and video. With the use of synthesizers, electronic information could be generated internally, and the information could be displayed optionally as audio and/or video. In the broader context of electronic culture, media-specific art practices are embedded in two parallel approaches toward technology: one being the incorporation and modification of existing instruments in electronic arts, such as radio frequencies and audio synthesizers, and the other the development of new devices, such as processors and other analog computers. Taken together, these approaches demonstrate that artists' curiosity with regard to emerging technologies is not satisfied by adding new tools or using them solely externally or as an extension to the artwork. On the contrary, whether the creative use of electronic tools that were designed for specific purposes, or the use of standard equipment against the grain, both determine the roots of electronic culture – open processes of experimental film installation, Fluxus, happenings, performance, sound and noise arts. All are set against the cultural dominant genres of painting, sculpture, music and film.

The major task of media tools seems to be to control the function and appearance of video: to manipulate, repeat and reposition the effects and build machines to systematize and maximize the possibilities of interconnection and modification – not unlike digital systems. In many applications of the 1970s and 1980s, the extended use of abstract waveforms in video goes hand in hand with extended feedback, which at the time was understood as creating a true videographic image, different from other types of media images. This difference was understood because of the abstraction from the pinhole perspective – video's permanent transformation and immediate presence as a constant flow of signals. In its circular structure, feedback can lead to the multiplication of an image's waveform, but it can also effect distortion, and it can be used in conjunction with other effects.

First of all, the invention of media tools was a dynamic phenomenon that amplified and liberated the flow of the electronic information from the constraints of the televisual broadcast standard. And that was a rather uncontrollable effect. When more programming functions were incorporated into electronic manipulation, the visual forms were maneuverable in all possible ways and finally entered the digital mode.

References

Angell, Callie (2002), 'Doubling the Screen: Andy Warhol's *Outer and Inner Space*', *Millennium Film Journal*, 38 (Spring), pp. 19–33.

Computer Museum (2012), 'Analog computers', *Computer Museum*, Amsterdam: University of Amsterdam, http://www.science.uva.nl/faculteit/museum/AnalogComputers.php. Accessed July 18, 2012.

Couchot, Edmond (1998), *La technologie dans l'art: De la photographie à la réalité virtuelle*, Nîmes: Jacqueline Chambon.

Cowell, Henry (1969), *New Musical Resources*, New York: Something Else Press.

Etra, Bill and Rutt, Steve (1973), 'Rutt/Etra Scan Processor (analog), 1973', in David Dunn, Woody Vasulka and Steina Vasulka (eds) (1992), *Eigenwelt der Apparatewelt: Pioneers of Electronic Art*, Santa Fe, NM: The Vasulkas; Linz, Austria: Ars Electronica, pp. 136-139.

Etra, Bill and Rutt, Steve (n.d.), 'R/E video synthesizer-operating manual' [unpublished manuscript], New York: Rutt Electrophysics, [n.d.] p. 7, Steina and Woody Vasulka Fonds, VAS B44-C3; Montreal: Daniel Langlois Foundation.

Flusser, Vilém (1985), *Ins Universum der technischen Bilder*, Göttingen: European Photography.

Gagnon, Jean (2006), 'A demo tape on how to play video on the violin', *Art Journal*, 65: 3, pp. 70–81.

Hagen, Charles (1978), 'A Syntax of Binary Images – An Interview with Woody Vasulka', *Afterimage*, 6: 1–2.

Hocking, Sherry Miller (1986), 'Rutt/Etra Notes on Development', Experimental Television Center website, http://www.experimentaltvcenter.org/ruttetra-notes-development. Accessed July 18, 2012.

Hocking, Sherry Miller and Brewster, Richard (1986), 'Image Processing', in David Dunn, Woody Vasulka, Steina Vasulka (eds) (1992), *Eigenwelt der Apparatewelt: Pioneers of Electronic Art*, Santa Fe, NM: The Vasulkas; Linz, Austria: Ars Electronica.

Hocking, Sherry Miller, Brewster, Richard, Bode, Peer, Rudolph, Hank and Schlanger, Matthew (1980a), 'Jones Sequencer in Experimental Television Center Studio System Manual', Experimental Television Center website, http://www.experimentaltvcenter.org/jones-sequencer-experimental-television-center-studio-system-manual. Accessed July 18, 2012.

—— (1980b), 'Jones Frame Buffer in Experimental Television Center Studio System Manual', Experimental Television Center website, http://www.experimentaltvcenter.org/jones-frame-buffer-experimental-television-center-studio-system-manual. Accessed July 18, 2012.

Klüver, Billy (1966), '9 Evenings: Theatre and Engineering' [program], October 13–23.

Klüver, Billy and Martin, Julie (2003), *E.A.T. - The Story of Experiments in Art and Technology* (ed. NTT InterCommunication Center), Tokyo: InterCommunication Center.

Paik, Nam June (1984), *Good Morning Mr Orwell: Transmitter and Receiver* (ed. DAAD), Berlin: daadgalerie.

Poissant, Louise (1997), *Dictionaire des Arts Médiatiques*, Sainte-Foy, Québec: Presses de l'Université du Québec.

Sandin, Dan (1973), 'In Consecration of a New Space: A Color Video Process' [information leaflet] in Phil Morton, Dan Sandin, and Jim Wiseman (eds), Steina and Woody Vasulka Fonds, VAS B41-C16, 26. 1; Montreal: The Daniel Langlois Foundation.

Sandin, Dan (2004), 'Notes on Works', written communication to the author, June.

Sandin, Daniel J., DeFanti, Tom, Kauffman, Lou and Spielmann, Yvonne (2006), 'The Artists and the Scientific Research Environment', *Leonardo*, 39: 3.

Sherman, Tom (2007), 'The Premature Birth of Video Art', Experimental Television Center website, http://www.experimentaltvcenter.org/sites/default/files/history/pdf/ShermanThePrematureBirthofVideoArt_2561.pdf. Accessed July 18, 2012.

Vasulka, Woody, Schier, Jeffrey and Moxon, Tom (1979), 'The Articulator Manual' [unpublished technical manual for the Digital Image Articulator], Steina and Woody Vasulka Fonds. VAS B63-C1; Montreal: The Daniel Langlois Foundation.

Youngblood, Gene (1989), 'Cinema and the Code', *Computer Art in Context: ACM SIGGRAPH'89 Art Show Catalog*, special issue of *Leonardo*.

Notes

1. The Rutt/Etra Video Synthesizer and the Rutt/Etra Scan Processor are used interchangeably for the same device.
2. Poissant quotes are translations from original text in French.
3. It has been argued that Paik could actually have recorded the first video with a portapak as early as 1965 (Sherman 2007). Clearly he could not have rewound the tape, but would have needed to use the industrial broadcast standard video recorder of a TV station. It is also established that before Paik recorded and presented his 'first' videotape in October 1965, Andy Warhol produced his first videotape in August 1965 with Norelco video equipment, which he included in his first double-screen film showing, *Outer and Inner Space* (1965), where Edie Sedgwick's image on video is presented in dialogue with her image on film (Angell 2002: 19–33).
4. For a closer examination of the Vasulkas' experiments with electronic sound-image manipulation, see Jean Gagnon (2006).

Analog Meets Digital In and Around the Experimental Television Center

Kathy High, Mona Jimenez and Dave Jones

Dave Jones was one of the key people who witnessed how computer processes were integrated into the artist studio at the Experimental Television Center (ETC), and he played a central role in designing and building elements of the system. Other designers/programmers/builders included Don McArthur, Walter Wright, Paul Davis, Richard Brewster, Peer Bode and Matthew Schlanger. Jones was exploring image capture in the early 1970s, which eventually led to his invention of the Jones Frame Buffer, as he explains in the first part of the interview. In these early days (ETC was founded in 1971), the computer was most successfully used in the studio as an accurate and repeatable source for control voltages, continuing with the Center's focus on using audio signals and waveforms to affect video. Image capture was a research goal, albeit an elusive one and not necessarily an end in itself. Rather, research with image capture was deeply connected to ETC's emphasis on experimentation in real time.

Interview with Dave Jones by Kathy High and Mona Jimenez, July 31, 2007

Mona Jimenez: I remember that you built your first A to D converter sometime in the early Seventies.

Dave Jones: I was designing my Jones Colorizer. And in the process of doing that, I decided it needed some sort of a keyer in it that would take the image, for anything above or below a certain shade of gray; it would clip that image. So I had worked out a fairly clean circuit for clipping the video image. In reading a little bit about conversion, A-to-D conversion and digital conversion in general electronics magazines and stuff, I realized that a series of these clipping circuits tied together would give me basically an A-to-D converter for video that would slice the image up into shades of gray and then turn it into a digital number, a multi-bit digital number. And then feeding it back into a D-to-A circuit would get it back to video. The electronics magazines I'd been reading were about audio conversion analog to digital, and were using much slower types of converters. There weren't really any single chips available at that point that would do video A-to-D conversion, although there were some that came out around that time and were extremely expensive.

MJ: What year would this have been?

DJ: '75. I think there was a chip that came out around that time, and it was 600-and-something dollars for the chip. There was no way I could buy that. So I was using these cheap little comparator chips and feeding them into a digital chip that combined the outputs of sixteen of them. And this was something that I'd developed at home on my own. It wasn't part of stuff at the TV Center.

And so I built this circuit that divided the image into 16 shades of gray, and then basically put it back together again. So you got back out the original video image, but now the shades of gray were stable. So everything between black and one-sixteenth of the image was now just black; and everything from that point to the next threshold was the next shade of gray. So it quantized the image, divided it into bands. And I played around with it a little bit, because between the A-to-D and the A, you had the four bits, where it was converted into a binary number; and by reversing those bits, you totally scrambled the shades of gray. So the 16 shades of gray were not consecutive black to white, they were dumping all over the place. So it was black and then middle gray, and then dark gray, and then almost white, and so on. So it gave this kind of zebra-striped appearance to an image. It ended up showing all sorts of little detail. Because in the original image, where you had minor steps from shade to shade, now you had major steps in the shades of gray, from shade to shade. So all the little detail, and all the noise and any other parts of the image now kind of stood out and were this kind of jumble of shades of gray.

So I played around with that idea a bit, and then decided that once it was a digital number, a four-bit number, that maybe I could feed it into a memory chip and store it and hold it for a period of time. Because all this conversion was real time, so there was a lot of detail in the image. It was digitized as far as shades of gray, but not as far as time.

So I got a very small memory chip, and I would just store the shade of gray for a few pixels, and then I would grab the next shade of gray, and then the next shade of gray. And I divided the image into 64 points across the line. So now the image was, like, 64 columns. And in each column, on each scan line, was a shade of gray for that portion of the scan line. So it had the full resolution vertically; but horizontally, it was divided into little segments. And then, by changing the way that the memory chip was timed, as it grabbed those 64 points across the line, I grabbed on one line and played it back for the next set of lines. And that divided the image into 64 x 64 blocks, where each block was then holding the shade of gray that came from the upper corner of that block. So it was a live image, but it was now heavily pixelated. It was now just this big, blocky image. And the circuit that did this was called a 'line buffer', that basically just stored one line of the image, 64 points on that line, repeated it for – I forget – 16 or eight lines, and then it grabbed the next line and repeated it again. So it created this kind of a big, blocky, pixel-looking image.

[...] In the beginning of '77, I moved to the Catskills, and Gary Hill and I shared a place there for a while. And the intention was, he was going to put me up and pay for the materials, and I was going to design some equipment that he could then have and use;

and at the same time, teach him how to duplicate the equipment, so he could make more copies and have big piles of equipment…

I sold him the A-to-D because I needed cash. And so I sold him the 16 shades of gray A to D. And then there was a surplus electronics company across the river. We were living in Barrytown, NY. And in Kingston was this surplus electronics place that had a big cardboard box, about this big, full of chips, most of them leftovers from projects at IBM, where they pulled parts and threw them in the box, and then they sold all their surplus stuff to this place. And I went in there one day and, for forty dollars, I bought probably 2000 used integrated circuits from these guys, and brought them back and started sorting through. And there were some memory chips in there. They weren't very *big* memory chips, but you know, they were more than 64 points; they were a 1000 points, and 2000, 4000 points, things like that.

So I built a circuit using those, to make a 64 x 64 frame buffer. So that instead of just capturing one line, it would capture 64 lines and grab an entire frame. And I played around with that a bit. Basically, you could digitize the image through it. And you would see live images at 64 x 64; and then at any point, you could hit the button and freeze an image, and it would hold it. And I believe the first time that I showed it to Gary, he wasn't there, and I pointed a camera at my face and froze the image of my face. And then I left and went somewhere else. And he came in later that afternoon into the studio, and there was my face frozen on the monitor. He reused that trick a few months later, when Steina [Vasulka] came to visit. And he froze his face on the monitor before bringing her back to the house to take a look at some equipment.

Kathy High: So this was actually a big deal. I mean, obviously.

DJ: Yeah, this was something that, you know…

KH: …was mind-boggling.

DJ: You would see it on TV, maybe, but it wasn't something that was readily available to the small-video person. So, you know, we were definitely blown away by the fact that it worked. And the A-to-D by itself, separate from the 64 x 64, by that point, was finely tuned to where it would actually – because it was so real time – pass color and rearrange the color. So besides turning out 16 shades of gray, you could feed a color image in and the subcarrier shape of the video would get chopped up and the shades rearranged, and now you had wild colors all over the place. Or depending on the shades of gray you chose, when you rearranged it you'd get subtle colors or strong colors or a mix of both.

I built a circuit in between the A-to-D and D-to-A called an ALU [arithmetic logic unit], which is a sort of a very simple mathematical digital chip used in digital processing. It basically adds two numbers, subtracts two numbers – very simple types of math type functions. And so I fed the A-to-D into one side of it and fed some switches into the other side of it, and then the output went to the D-to-A. And by changing the switches, you could change how the shades of gray got rearranged. And so you could end up with a positive, a negative, or just totally scrambled shades of gray.

And when you did that with a color image, or even just a subtle amount of color mixed in with a black-and-white image, you would end up with all of these different colors instead of just a shade of gray, but still in this kind of quantized look. So you'd have large areas of an image, and they would all be this, like pastel – kind of a beige or something – and then another shade of gray next to that might be a bright red. So it was some fun stuff. Gary did a videotape, I think called *Bathing*...that used that A-to-D and ALU unit. (See Color Plate 23.)

Spatial and Intensity Digitizer

Accounts vary on what Dr. Donald E. McArthur intended to invent when he designed the Spatial and Intensity Digitizer (SAID) in Binghamton, NY, in 1974 or 1975. Some say he was designing a time base corrector,[1] a highly desirable broadcast device for the stabilization of signals from various low-end video sources; a device that was too expensive for an artist's space like the ETC to acquire. Design engineer Jeffrey Schier recalls that McArthur was trying to design a low-cost analog-to-digital converter for video (Dunn 1992: 143).

A rather obtuse description from ETC's 1975 application to the National Endowment for the Arts suggests that the SAID did more than act as an analog to digital convertor: 'The SAID reproduced a black-and-white camera image which is constructed from memory by horizontal and vertical structures varying in gray level which are controllable' (ETC 1975b: 1). ETC's Studio Manager, Hank Rudolph, remembers that the excitement of the SAID was in seeing the whole cycle – analog to digital and back to analog – rather than the fact that it could convert an analog signal into digital form. The SAID did not store bits; rather in Rudolph's words it would 'signify digitization' (Rudolph 2010: 26). Woody Vasulka calls the SAID a 'moment of change in an esthetic norm in one's mind [...] looking at a digital image broken down into the numbers and reassembled again in real time' (Dunn 1992: 142).

The presence of McArthur, who was a visiting professor at the State University of New York (SUNY) at Binghamton, brought new skills to the Center in the early days of ETC's engagement with digital devices. In a 1974 ETC funding proposal to the New York State Council on the Arts (NYSCA) for 'Systems Approach to Video Art' MacArthur is described as key to 'the development of a system for synthesizing, processing and controlling video images with greater flexibility, reproducibility and precision than is presently possible' (ETC 1974a: 1).

The SAID was installed in the ETC studio for a number of years, but little documentation remains of the machine. McArthur later collaborated with designer Jeffrey Schier to create the McArthur/Schier Digital Image Generator for Steina and Woody Vasulka.

MJ: Were you in contact with other people who were trying to build A-to-D [devices], line buffers and frame buffers?

DJ: I don't remember. Around that same time, people were discovering computers. And so there was that side of digital that people were discovering, but it wasn't for image generating.

MJ: What would they have been discovering about computers at that point?

DJ: It was *very* simplistic kind of stuff. Chuck Kennedy [of the Videofreex] got a hold of a computer. And I remember sitting there watching him having to key programs in one byte at a time on switches on the front panel, where he would have a reference sheet, and he would have to put these eight switches in a specific pattern, and then hit a button, and that stored one byte. And then he would change the switches to another pattern, hit the button; that would store the next byte. And he would have to do that 200, 300, 500 times to get a very small, simple little program in the computer, before it would actually run anything.

MJ: And that wasn't necessarily running graphics.

DJ: No, I mean, those were running very simple things. In some cases, they were maybe putting out a square wave that would be used like an oscillator. And there were some plug-in cards that some people had together around that time, I think around '76 or '77, that were very simple kind of graphics cards…character generator type stuff, and very crude patterns of images. But that assumed that you managed to get a hold of one of those kinds of video cards, which were fairly rare…

MJ: So I'm hearing three different things. One is your work with the capture of a frame, which is a very big deal. The other is generating computer graphics that could be used with video. But I also understand there was computer control of video. Are those three of the main themes that were [present at the time]?

DJ: The first computer the TV Center got, which was also in '76 or '77 – I think it might've been '77 – was the LSI-11, which was a single-board version of the DEC PDP-11. And its basic function was recording and playing back control voltages. It wasn't really capable of making video. But because a lot of things in the studio at that point were control voltage – I built my colorizer all with control voltages; the SEG [special effects generator] had a control voltage input for adjusting of the wipes; the Wobbulator could take oscillators. So there were a number of things that had control voltages…

And so the computer was set up with some voltage inputs and a number of voltage outputs, and some very simple software that was able to record and edit patterns of voltages. So you could – instead of just playing an oscillator that was one repeating shape like a triangle wave or a sine wave – you could make a much more complicated pattern of voltages going up and down, and then play them back from the computer. And you could speed that up and slow that down. And that was the primary function of the LSI-11 at the studio. It was basically not making video, but making control voltages and being able to play them back, and feed a control voltage in and record it, play it back faster or slower; or define a pattern and then play that pattern back.

MJ: I understand that part of the introduction of the LSI-11 into the Center was not just to put the computer in there, but also provide an interface that was useful to the artists, so they could deal with it on a knob kind of level, on a physical level versus a programming level. Is that right?

DJ: Right. We were writing the software that would make it work; and then once it was up and running, [artists] would be able to just run the program and throw a couple of switches or turn knobs, and be able to play back the control-voltage sequences. So they would be able to go into record mode and turn a knob in various patterns, and then throw the switch back and that pattern would come out through the control-voltage output.

MJ: So would you be able to actually send a pattern or a sequence of control voltages, or just individual control voltages, individual patterns?

DJ: Well, each output, you could adjust the waveshape, basically, and record a voltage pattern. And I can't remember how many outputs the LSI-11 had, but it had multiple control-voltage outputs. So you could have several outputs happening at the same time, going to different places in the system. It wasn't particularly easy to use. The software was fairly crude. The computer itself was a real pain to get booted in the first place. And so there were probably not a lot of people that used it. Probably only a handful really did anything with it…

DJ: I would come back to Binghamton every week or two and check in, and write a little code, and you know, see what the project was up to. At that point, Paul Davis was doing a lot of the work on the [LSI-11]. He did a lot of the assembly. Walter Wright was there doing a lot of the layout and assembly.[2] And I was just coming in and helping out and throwing my two cents in every once in a while.

It started in '77, went through, like, '78, '79, around in that range. The TV Center moved [from Binghamton] to Owego in the middle of '79. And I don't really remember the [LSI-11] computer surviving much beyond that, as far as being used. Sometime around '80 or '81, Ralph said to Paul Davis, 'We need to go to the next step here and, you know, get something a little more useful.' Plus, [Ralph] wanted something for his own use at home in his studio, which he had been building up.

And so Paul put together a computer system similar to something he'd been working with at [SUNY] Binghamton, which was a Cromemco Z-2 computer. This was a kind of a more modern microcomputer. The LSI-11 was made by one of these big companies that made computers for Boeing and all these other kind of places; whereas the Cromemco was kind of part of the small microcomputer generation and revolution that started in the late Seventies. It was the next step beyond the kind that Chuck Kennedy had, with the switches on the front, where you had to really work at it to get a simple program in. This one actually had disk drives. And it had a graphics card and it had serial ports and...you know, it had most of the kind of things that a normal computer these days has. It had an operating system called CP/M that was becoming very popular back then. And there was a plug-in card for it that was available, which we got, called the D+7A, which was digital plus seven analog. And it allowed you to feed analog voltages out and have switches that went in, analog voltages that would go in, and digital pulses that would come out. So it basically was switches and knobs going in and pulses and analog voltages coming out.

And so we used that and made a little box to interface with it and bring it out to jacks and knobs that we could turn and interface into the system. And then started writing different types of programs to work with it to do things, to feed voltages out and to record them – similar to what the LSI-11 had done. It was much easier to write programs. So we ended up doing many more variations of programs using it.

And then at some point around then, once we started doing that, Ralph came across a video frame buffer board from a company out in California that would plug into this computer. And it would capture frames of video into the computer. And so we started writing programs that would control that board using the knobs from the interface box. So you could adjust the speed of how often it captured an image or how fast it played back, and slow it down, speed it up. You could use it to adjust the resolution of the captured image. Things like that. So [the interface] became...physical knobs to control the software that was controlling the hardware inside the computer.

MJ: So was another difference between the LSI and this Cromemco that the graphics card would enable the storing and playing back of images, versus just control voltages?

DJ: Yeah. There was a graphics card, which was in the computer for going on the screen, just so you could see what the computer was doing. But then there was also the frame buffer card. And the frame buffer card was something that had video in and video out, and could actually capture frames of video and put them back out. The computer graphics card wasn't something that had a video output; it just was something that stored and displayed the characters when you type at the keyboard and that kind of stuff. It was the computer screen for controlling it.

And actually, now that I think of it, there was something in between those two. After we got the Cromemco – I don't remember if we got this right away, or fairly shortly afterwards – there was a plug-in board for it called the Dazzler. The Dazzler was a graphics type of a

Language (Software) Development

> I think software must be concerned with composition, first of all. First must be the elements and attributes of design. Second, the software must be capable of both analyzing and synthesizing images. Third, the software should have a mind of its own. It should be capable of reprogramming the synthesizer and reprogramming itself. Fourth, it must interact in real time with the artist. And fifth, it has to cost almost nothing. (Wright 1977: 6)

ETC staff and collaborators used programming, as David Jones notes, as soon as computers entered the picture. What is surprising is the extent to which software development was part of ETC's early years. In 1974, ETC submitted its first request for software development to NYSCA as part of a proposal for an artist's residency for Walter Wright. Wright's activities were intended to expand ETC's workshops and studio facilities to interface between computers and other video instruments. The plan was for Wright's software programs to be accessible through an interface external to the computer and more akin to the knobs and switches video artists were used to, as opposed to through the more indirect and non-intuitive command line interface coupled with a keyboard. (See also below 'Interfaces for Computers: The D+7A Box'.)

The computer would allow the artist to make quick and automated moves that were impossible with human hands:

> The computer will be used to preprogram a series of images and transformations which would be impossible to achieve with manual controls. The programs can be edited and combined to build completed scores for the synthesizer. Programs will be stored on audiocassettes [...]. The ability to program will require the development of a special language for the computer and a visual notation system for composition. (ETC 1974b: 2)

There were six areas in which Wright planned to develop software programs. Interpreter was to translate between human recognizable commands (a score) and machine language; Interactive Monitor would allow for playback and editing of the score; Timing was to maintain a frame count; while a Control Program would handle the sequence for analog-to-digital conversion. Converter would have translated back from machine language to a form that the user could update and edit, and Composition would have allowed for scores on larger computer systems and would have automated and organized images or scores.

Wright's residency was intended to create software and hardware and to create documentation to be distributed at cost to the public. While Wright never created distributable software, he did report on his research, including programming he was working on with Woody Vasulka, at the 'Design/Electronic Arts' conference that was held in Buffalo, NY, in March 1977. His focus was on the computer as an external control to video synthesizers, planning for 'predictable unpredictability'. He explains: 'You may not be able to predict exactly what each image will look like, but you will be able to predict certain properties of them [...] the artist interacts most effectively with the system by controlling those probabilities in real time' (Vasulka 1977: 8).

Request for Assistance
Experimental Television Center Ltd.
Program: Artist in Residence, Walter Wright

Attachment #2
Notes page 2

Timing program maintains a frame count. Vd/2 is used to interrupt the Monitor and allow the program to increment a preassigned word in word in memory containing the frame count. Control then passes back to the monitor.

Block diagram of programs

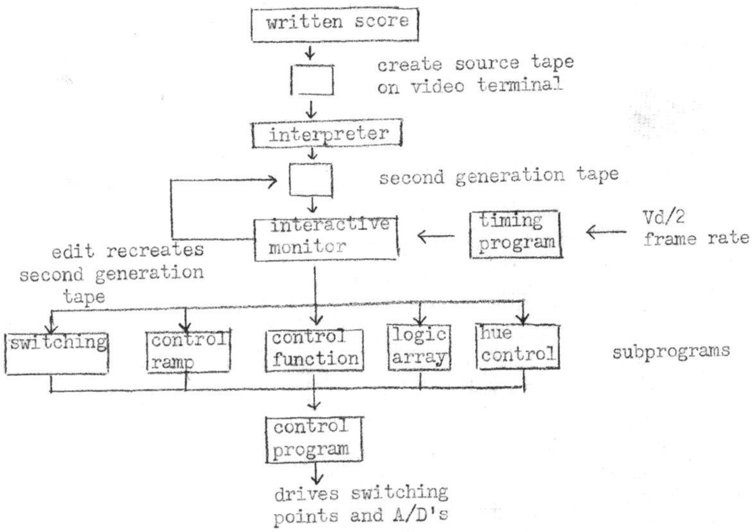

Control program drives the analog to digital converters and switching points. The program outputs an address which specifies a particular A/D or bank of eight switches and a data word indicating voltage level or on/off settings.

Converter is a program which creates a new source tape from a fully edited second generation tape. This is necessary because the second generation is in machine language and must be reformatted as commands for the user to understand.

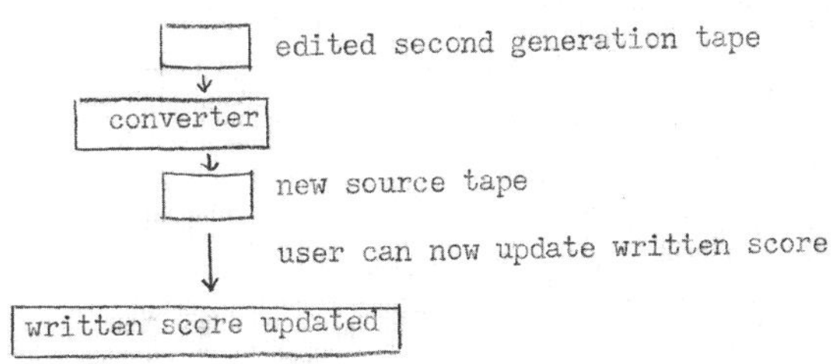

Figure 1. A page of ETC's application to the New York State Council on the Arts for Walter Wright's software development project (1974). (courtesy. Walter Wright and the Experimental Television Center).

card, not very high resolution, but easily controllable, and it put out standard video. So it had a recordable output, which the normal character generator–type graphics card did not have. And the Dazzler could display maybe 64 x 64 graphics. And so we were able to create patterns on this, and put those patterns out as standard video and be able to record them and feed them into the system. It didn't have any way of really capturing video, but it was basically a pattern generator. It was for making colored patterns and squares.

And I remember writing some software where you'd turn the knobs on the analog box to adjust the kind of oscillations moving through the Dazzler to make patterns. And the Dazzler had enough memory in it that you could do sequences. So you could store a pattern and then bring up another page with a different pattern, and then another page with a different pattern. And so it was possible to sequence the different frames in the Dazzler while you were drawing on each of the frames, and create moving patterns on the screen. As they sequenced, the pattern would repeat as a moving pattern.

Then we got the [Cromemco] frame buffer, which had much higher resolution and did real video in and out. And it was, I think, 256 x 256. And I remember capturing some images with that, and then through software, reducing the resolution and putting them into the Dazzler. Because the frame buffer would capture a frame and play a frame. But because the Dazzler had this low-res repetition of multiple frames, we could capture images in the frame buffer and then put those images in the Dazzler and, in low-res, sequence the images and get this kind of motion playback through that. So it was similar to the resolution of the buffer that I made at Gary's place in the Catskills, except this was color. And because it had sixteen frames, you could play these patterns and repeat motion and stuff. That kind of led me to build my next frame buffer with multiple frames.

MJ: And that would've been what year, you started working on your frame buffer?

DJ: I think in '79, I built the next version up from what I had made at Gary's place. Peer [Bode] supplied the money to buy the parts, and we did a 256 x 256 single-frame frame buffer. So it was much better resolution than the 64 x 64 that was at Gary's.

MJ: So you were capturing, then, these single frames, and then you were trying to play them back at faster and faster rates over time?

DJ: Well, you couldn't really – I mean, it was basically about capturing single frames or keying a portion of an image onto that single frame. So we would cut out some portion of the image using a key clip, and the static frame would then have this image moving over it. But it was a single frame. It didn't have multiple frames, so there was no repeating motion. There was no sequencing that happened of time capture.

It was just building up kind of a composite image by layering parts of the live image on top of the stored image. And then from there, we decided that multiple frames were definitely needed. And the TV Center came up with the funding to do the frame

Interfaces for Computers: The D+7A Box

> I just can't do mathematics…I said, 'I do not want to sit at a keyboard and punch an H or something, and have it mean something else up there on the screen. Give me something to get a hold of. At least let me turn a knob or throw a switch.' That's where the D+7A Box came from, that push when we first started playing with computers. I wouldn't sit with keyboards and learn a language because with language, I'm right back to mathematics. I can't do languages either. 'Qué pasa?' is about as far as I get. (Hocking 2005: 9)

As early as 1974, ETC had articulated the importance of a useable interface between computers and image-processing tools in funding proposals to NYSCA. Also, through Sunday meetings where he cooked and worked the philosophy 'keep 'em high, keep 'em happy', Hocking started a conversation about how to make computers part of the ETC toolset (Hocking 2005). Paul Davis, who worked with Hocking at SUNY Binghamton, became an integral part of the Center. Around 1979–80, ETC acquired the Cromemco computer with a Z-80 microprocessor and several commercial add-ons; a 'CAT' Buffer that could capture a single field of video at 256 x 256 pixels and 16 colors; and a D+7A interface box (ETC 1997).

The D+7A box had seven channels for analog-to-digital and digital-to-analog conversion but a nonintuitive interface. The interface provided switches, potentiometers and input jacks for ease of use.

Signals were brought into the computer from a Jones Keyer, a Jones Colorizer or a special effects generator. The user had to boot the program using the keyboard and choose a program to run, but then the interface bypassed the normal command line interface to enable the user to change parameters manually. In other words, the D+7A box gave artists an interface that resembled those of analog devices in the system. An ETC-created reference, 'The Digital Imaging System' gave instructions to users (ETC 1977b). The concept was repeating in numerous devices, such as the General Purpose Interface Board, created with NYCSA funds in 1983.

buffer project. And so that was the same 256 x 256 converter, but with multiple frames. And that used a memory card that had 16 frames per memory card. And so if you had multiple memory cards, you had multiples of 16. 16, 32, 64 frames. And then you got *real* motion. And that's what we've had at the TV Center since around that time, since the mid-Eighties. I think we proposed it and tried to get the funding around '82. And we finally got the funding and finally actually built the boards around '85 for that.

MJ: So your first version, the one that captured single frames, you called the Jones Buffer, as well?

DJ: Well, I don't think – I mean, Ralph is the one who added my name onto the ends of all the machines that I built. And, you know, the 64 x 64 one was just the 64 x 64 frame grabber. The other one was the 256 x 256 frame grabber, or Peer's frame grabber. And then by the time we etched the boards and started making multiples of the final one in the Eighties, Ralph had decided that was the Jones Buffer. So like the Jones Colorizer, it got my name tacked onto it.

MJ: Good idea.

DJ: Yeah.

MJ: So I think I read somewhere that you bought your first desktop computer in 1985. And I was wondering when the Amiga [computer] started to appear, and how that got integrated. Was it around the same time, the mid-Eighties?

DJ: It was in the mid-Eighties. My first computer that I bought was a PC clone, which I bought because I had seen some computer-generated printouts to make circuit boards in the mid-Eighties at Alfred University. And actually, one of the boards that I made at that

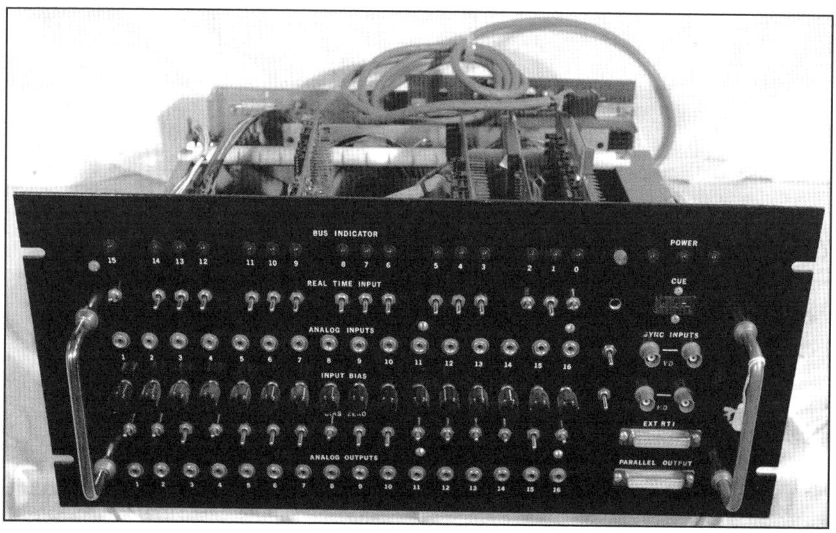

Figure 2. A 16-channel interface box for the LSI-11 computer designed by Paul Davis and built by Paul Davis and Richard Brewster (c. 1977). (photo. Sherry Miller Hocking. courtesy. Experimental Television Center).

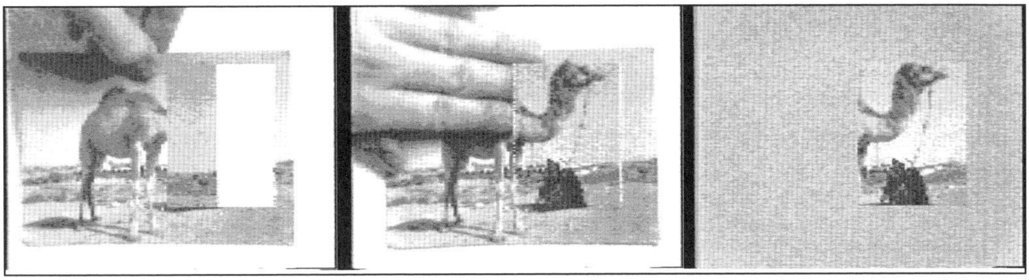

Figure 3. Three stills from *Camel with Window Memory* by Peer Bode (1983). The tape was made using a frame buffer built by Bode and Dave Jones that allowed capture of one still zone and one live zone, creating the illusion that the still zone (the small rectangle) is in front of the live image of the postcard. (courtesy. Peer Bode and the Experimental Television Center).

time was a printout on their big CAD system. And so I figured, 'Well, that's kind of the way things are going; I need a CAD system'. So I bought a PC clone, and I bought one of the first versions of AutoCAD that came out back then, and started designing my circuit boards on the computer and printing them out on a little pen plotter, using felt pens; and then taking those pieces of paper down to a photo lab in Binghamton, and they would photograph them into double-sized lithos; and then I would send them to a circuit board house to have them made as circuit boards. I think that was around '85, when I got that machine.

I don't remember exactly what month the Amiga became available. I think I got mine about '86. Like the end of '86, something like that. There were a few people that had started to get them and were playing with them. And the graphics on them were incredible for the day. And so I couldn't afford to buy a complete system, but I met somebody who wanted a monitor, and I wanted the computer. Since it put out normal video, I could use my own video monitor for the computer monitor. And so I bought the complete system and sold him the monitor, and kept the computer part and started playing around with that.

MJ: So obviously, the Amiga offered the paintbox functions…

DJ: It had animation. Over time, it added a lot of other things.

MJ: And digitizing…

DJ: It had sound cards, it had video capture, and it had a genlock. There was 3D software made for it. There was animation software made for it, both vector-based animation, as well as frame-based animation.

MJ: So that was kind of a shift in the way computers were used at the Center? And also by people like you?

DJ: Well, no, there probably was one fairly early at the TV Center. But it was separate. I mean, the Cromemco was still used at the Center for a long time. And the Amiga didn't replace that. The Amiga was all about doing graphics and doing colored shapes moving around and stuff like that. I can't remember when the Center got its first Amiga, but it was maybe a year or two after that. It was like '87, '88, I'm guessing.[3] But I'm not really sure. And I think the first uses were basically flying triangles and ovals and things like that moving around, and then moving on to capturing video and doing sequences of video and stuff like that. So [the Amiga] did eventually replace the frame buffer part of the Cromemco, because the frame buffer was always kind of cranky and had loose wires, and never was really 100 per cent predictable. Would work great for a while, and then it would start screwing up, and you'd have to go and wiggle wires and move connectors around. And then you'd get it working again, and you know, a few months later, then something would come loose again.

MJ: So could you talk about the similarities and differences between the [earlier] frame buffer, your Jones Frame Buffer and the image capture parts of the Amiga? Were they kind of just several different tools that did the same thing?

DJ: Well, they were fairly different. The Amiga was actually a color device. You know, my frame buffer was never a color device. It was black-and-white in, initially black-and-white out, although I did make a colorizing type of a card for it, a color map card that assigned colors to the different shades of gray. So you had an alternate output that was like quantized color.

MJ: When would that have been?

DJ: It was done around the same time. By the time we finished the main boards, I think that board came months later. So it was probably, you know, around '86, something like that. Before I got my Amiga. And there were some other kind of computers and graphics systems that were happening around that time. Like the Z-GRASS system, through Tom DeFanti and the people in Chicago. Which was basically a Bally video arcade game that had special software running on it.

MJ: I wanted to actually ask you about that, because it seemed like [the Z-GRASS] was really a graphical device…

DJ: Right.

MJ: As opposed to the initial interest by the Center more toward control voltages, controlling video. [ETC was] less about development of graphical systems but also had an interest in frame grabbing and capture.

DJ: Well, it was kind of an evolution of what the computer was capable of. The early computers weren't capable of grabbing frames or putting out video. But they were capable of manipulating voltages. And then the second generation ones started with the control voltages, because since it worked on the first one, that was where we obviously kind of saw it going, to begin with. But then as different add-on cards became available for the second computer, then we added those features to it. So it then became much more about being able to make video with it.

MJ: So what about the vector graphics? Was it that the vector systems were just not affordable, but they were perhaps in Chicago because it was the university lab that [Tom DeFanti and Dan Sandin] were using? Or do you think there just happened to be different personalities interested in different [things]?

The Center for Microprocessor Research

ETC envisioned a Center for Microprocessor Research that would study software and hardware development for microprocessors to enable interfacing with video and audio synthesizers, including 'consulting with artists and arts organizations concerning the construction of small-format computer systems and their applications in video and other arts' and 'workshops on theoretical and practical aspects of the design, construction and uses of microprocessors' (ETC 1977a: 1).

In addition to enabling artists to integrate video, audio and computer tools, the ambitious project would have produced software and hardware to give control over '(2) lighting systems for use by performance centers (3) control over animation processes in filming and direct generation of images on film (4) control over mechanical devices for application in sculptural and environmental works' (ETC 1977a: 1).

Areas of research included a 'programmable filter bank for image modification and enhancement', a frame buffer, a color encoder, 'pseudo content addressable memory with variable drift', character generation and a 'raster partitioning system' (ETC 1977a: 2).

Periodically ETC pitched other concepts that went unrealized. In 1975, ETC applied to NYSCA for a 'Technical Innovation Project' to build 'a Digital Raster Manipulation (video storage unit) that combined frame storage with moving sections of the raster, collapsing and reversing the raster, and enlarging and reducing, combining aspects of the Raster Manipulation Unit and the Rutt/Etra Video Synthesizer' (ETC 1975c: 1).

DJ: You mean full vector systems?

MJ: Well, what they were doing with the Z-GRASS.

DJ: Because they were developing that based on, I think, some commercial input that they had had and projects that they were working on, and [the Z-GRASS] was kind of applying it to a small computer to make similar kinds of things happen. Whereas we didn't really have that commercial input, influence. We were just kind of coming up with things on our own or adapting what we could find off the shelf…to do something different.

MJ: So it was probably more serendipitous and less philosophical.

DJ: Yeah. I think it was just a matter of situation and influences that were around the different groups of people. What resources they had available to them and, you know, directions they got pushed in by the places that were providing funding.

MJ: Right.

DJ: And we were mostly funded by arts organizations and by ourselves…

MJ: And the [New York State Council on the Arts], which was open to development of tools.

DJ: Right. And so basically, we'd throw out ideas, and once in a while they would say OK. We threw a few out that they didn't say OK, so you know… It happened both ways.

MJ: Eventually, I guess, the control voltage part of computers just kind of faded away.

DJ: As computers progressed, things like MIDI came out, which were really designed around getting music, and then control voltages for audio synthesizers. Or just the types of audio synthesizer devices and modules that were available really could provide a lot of the same kind of functions that we had been trying to do with the computer with control voltages. So it really ultimately came down to [the computer] not being as useful for doing the control voltages, just because there were other ways of doing it. It wasn't that the computer wasn't good at it, but there were other devices that would do that same kind of thing.

MJ: And they were integrated into the system at the Center.

DJ: Eventually, yeah. I mean, as we added different parts of the system. We had our own control voltage boxes that we did a lot with. And then we added various outside commercial synthesizers, like the Korg control voltage processing and audio-processing box. Stuff like that. And eventually, some MIDI interfaces that were control voltage in and out to MIDI. So you know, the computer became less targeted towards control voltages, and it made much more interesting video images.

MJ: You did the frame buffer and were still working on kind of custom tools, even though there had been this shift, where there were desktop computers and the Amiga with improved graphics and interfacing with video. Were there still custom devices that were being built? Or was there really no need for that at that point?

DJ: There were, but they were kind of shifting direction. You weren't really seeing much in the way of, like, analog video processing or anything like that coming out as custom devices. You were seeing people doing stand-alone devices for a specific application or for a specific artwork. There were some people that built video frame buffers and colorizer-type circuitry – or not really colorizers, but sort of that same color look-up type stuff that I was talking about doing with the color outputs of the black-and-white buffer. But building that type of thing into a box, and then selling that as part of an artwork, so that the artwork had this kind of black box that the video camera went through and created the effect with whatever was happening on the camera.

MJ: So was it more at a software level, then, than a hardware level?

DJ: Well, at that point, it was still hardware, because software wasn't fast. It's only been very recently that software can go fast enough to actually create video.

KH: Recently, as of?

DJ: Well, I'd say the last five, six years; it's been fast enough. And it's only even the last couple years where it's fast enough to really go at full speed. Even five years ago, you could get video images out, but if you did much processing to them, you were down to 15 frames or ten frames a second. Whereas now the computers are fast enough you can get a full 30 frames a second, and still process the image a reasonable amount in real time, without losing the quality or giving up the frame rate. And the computers are still escalating in speed, you know, year-by-year. So now you're getting to the point where you could be doing HD or doing multiple channels of video, and still pretty much keeping up with the full video rate going through it. But, you know, up until I would say the early twenty-first century, the computer could only really control hardware that would process video. It couldn't really do the video in software directly. It couldn't create the pixels in real time, or you know, not at full resolution, full speed.

MJ: Right. So I guess I'm trying to link what was happening at the Center [in the twentieth century] – if there is a link – to some of the tool development now. There's a lot of software development, perhaps some hardware as well.

DJ: Yeah, I don't think there's a direct link, in a way, but there's a lot of influence. I think that a lot of people that were developing software for doing image manipulation saw things that came out of the Center, or saw things that came out of other places like that, and that was their sense: What do we want to make the computer do in processing an image? Because, you know, I mean, a lot of them, the software begins by emulating your classic SEG. And then when the designer of the software sees something more complicated coming from somewhere else, then they start adding that. A lot of the software development hasn't been totally out-of-the-blue development, you know, based on [the designer's] own pure concepts. Some of it has, but I would say the majority of it hasn't. Most of it is mimicking things that they've seen somewhere else.

MJ: People that are related to the Center? Or are you just saying in general?

DJ: No, I'm talking in general, with software that's being developed all around the world that's being used to create or process video images. And you could even see it in the Eighties, when things like the Video Toaster were being created. And the capabilities of the Video Toaster were based very largely on your classic SEG from a TV studio, because that's what the designers had seen. And so that's where they got their ideas, and that's what they then created, a kind of a new version of it. And then as they saw other types of devices and other images that were created, that came out of places like the TV Center, then those designers added features like that to their software and hardware. And the same thing's happening now with the pure software version of image manipulation. They're seeing old videotapes, or current videotapes, being processed with analog control-voltage type video stuff, and then they're building those types of effects into their software.

MJ: Right, MSP and Jitter.

DJ: Yeah, Jitter. And some of the overlays and processes there, some of them are taken from kind of classic video synthesizer-types of effects. They each design a little module to do some core effect, and then that adds to the features of the whole system. But a lot of the initial effects were just a basic keying or colorizing, and then they've grown from there. So I think that it isn't that the designers of that kind of software ever went to the TV Center, but I'm fairly certain that they saw tapes that came out of the TV Center, or places like the TV Center.

References

Dunn, David (ed.) (1992), *Eigenwelt der Apparatewelt: Pioneers of Electronic Art*, Linz, Austria: Ars Electronica.

Experimental Television Center (ETC) (1974a), 'Request for Assistance: Systems Approach to Video Art', Owego, NY: Experimental Television Center Archives.

——— (1974b), 'Request for Assistance, Experimental Television Center Ltd, Program: Artist in Residence Walter Wright', Owego, NY: Experimental Television Center Archives.

——— (1975a), 'Communication between ETC and Lydia Silman, Program Associate', Video History Project, New York State Council on the Arts, April 18th, Owego, NY: Experimental Television Center Archives.

——— (1975b), 'Computer Interface with Video Synthesizer', Owego, NY: Experimental Television Center Archives.

——— (1975c), 'Program Request Sheet', Technical Innovations Projects: Dave Jones', Owego, NY: Experimental Television Center Archives.

——— (1977a), 'Program Request Sheet: Center for Microprocessor Research', Owego, NY: Experimental Television Center Archives.

——— (1977b), 'The Digital Imaging System', Owego, NY: Experimental Television Center Archives.

——— (1997), 'Program Notes: 25th Anniversary of the Experimental Television Center', Owego, NY: Experimental Television Center Archives.

Hocking, Ralph (2005), 'Interview with Ralph Hocking by Carolyn Tennant', Owego, NY: Experimental Television Center Archives.

Jones, Dave (2007), 'Interview with Kathy High and Mona Jimenez', July 31, Owego, NY: Experimental Television Center Archives.

Rudolph, Hank (2010), 'Interview with Hank Rudolph by Kathy High and Mona Jimenez', July 31, Owego, NY: Experimental Television Center Archives.

Vasulka, Woody (1977), 'Design/Electronic Arts' [conference transcript], March 10–13 Buffalo, NY.

Wright, Walter (1977), 'Design/Electronic Arts' [conference transcript], March 10–13 Buffalo, NY.

Notes

1. In correspondence with NYSCA, ETC wrote that McArthur's 'original interest was to build a low-cost time base corrector using a theoretical approach.' The letter also refers to a videotape produced using the SAID. See Experimental Television Center (ETC) (1975a).
2. Paul Davis was the director of the student computer lab and an instructor at the School for Advanced Technology at the State University of New York (SUNY) at Binghamton. Artist Walter Wright was involved with the Center on many projects, including hardware and software development and education.
3. ETC had an Amiga computer in the studio by the fiscal year 1984–85 (ETC 1997).

Multi-tracking Control Voltages: HARPO

Carl Geiger and Mona Jimenez

Carl Geiger received arts funding for three years running (1974–77) for 'technical development' and 'computer-video research'. With one of the grants, Geiger hired Rod Fountain to write HARPO, a computer language intended for distribution among artists with interests in video and image processing. Geiger had already built a video synthesizer: 'my electronic imaging system [...] had parts of the Sandin Image Processor, parts of the Rutt/Etra [Video Synthesizer], and parts of various ideas, but it was never particularly one kind' (Geiger 2011).

While Fountain wrote the software, Geiger built a custom interface to the computer 'with knobs that were big and nice and sweet and smooth' (Geiger 2011). The artist could send impulses to an Altair computer using the knobs or a joystick. The computer would record the 'commands' on a 5¼" floppy for recall later. The real value of HARPO was in the ability to record multiple 'tracks' called 'SCORES', and to be able to edit and store them in the computer's memory.

While the idea of using programming to record and play back control voltages was circulating in the media arts community (see 'On Voltage Control' and 'Analog Meets Digital In and Around the Experimental Television Center' in this book), Geiger managed to produce a finished product complete with a manual. Geiger had the best of intentions for wide distribution, but in practice the software was used locally in the Syracuse, New York area.

Interview with Carl Geiger by Mona Jimenez, February 5, 2011

Carl Geiger: [With my synthesizer], so often you would just have an idea and patch it together and see what it looked like and turn the dials, and say 'Oh, that's interesting,' or say 'Oh, that's terrible.'

Mona Jimenez: So it was a real-time experience.

CG: Yeah, it was very interactive. You'd immediately get feedback, but then it also was terrible because the next day you'd say, 'Oh, that was really great, I've got to show that to Frank', and then you couldn't get back there. So in time what happened was it got so complex that you are looking at an image – perhaps I had a person in front of a camera and I was trying to do something – and you are trying to turn these knobs, and you just couldn't turn enough knobs accurately enough to make things work. So what I did was I got a grant from the National Endowment for the Arts to build a computer system that could record control voltages, in effect record the turning of a knob over time. And then you could play the recently recorded control voltages... and then do another. Adjust the control voltages over time while the other ones played back. So it was multi-track – it was layer upon layer of control voltages.

MJ: I hate to repeat [the word] control again, but could you then control use of the oscillators, the use of the control voltages, in a repeatable way?

CG: Yeah, the idea was that I could have an idea of something, a thing that I wanted to do – but I couldn't physically, accurately, turn the knobs in the right sequence. Over ten seconds, or over one minute, or over ten minutes. I couldn't physically turn eight different knobs accurately in any way, shape or form, so [HARPO] allowed me to have an idea and in effect do just one little part of it. Play through, record it, play it back, and then do the next knob while the computer played back what I had done previously and recorded a new version. I could do layering and then, in effect, could very accurately get repeatable results of the machine. You could say, 'I want this to transform from this shape into this shape and these colors to shift this way,' and all of a sudden it would jump over to this effect that looks like snow or whatever. You could make things over time that were repeatable.

MJ: So when you wrote HARPO, what operating system and platform... what computer were you using?

CG: So, this is the Altair Computer...to get this computer started you had to put in 23 bytes of information and then you would hit RUN and it would be smart enough to load its operating system off an audio cassette. Hit RUN, then you push PLAY on the audiocassette, and it would go and would be able to start the operating system. And then after a while I traded my senior year at Syracuse University for this dual 5¼" inch floppy disk drive. This thing held 125K of information and it was a godsend. Because the computer would start right off of this.

MJ: So you set up a keyboard interface to the synthesizer?

CG: Yeah, the synthesizer had a keyboard. The keyboard just generates control voltages, and so with the control voltages you could patch into anything that related to the image. And that's the whole beauty of this whole synthesizer because you could patch anything into anything, so you could do unexpected and unplanned and untraditional things.

MJ: That is one of the interesting things that I have been finding out in talking to people [for this project]. Really [in the 1970s] it wasn't about analog-to-digital conversion; it was about analog to digital to analog.

A>TYPE B:HARPO.DOC
'HARPO DESCRIPTION AND OPERATING INSTRUCTIONS'
First four letters of this file's name: HARD.
Dated: 2/21/79.
Author: Rod.

HARPO OPERATING INSTRUCTIONS

1. WHAT HARPO WAS DESIGNED TO DO

The basic idea with HARPO was to allow someone with only a marginal knowledge of computer concepts to control video effects in an accurate, fast, and repeatable manner. In reality the user probably needs either much coaching or much knowledge of computer concepts to suceed. The video effects are created by sequences of control voltages applied at specified time intervals to video hardware through digital to analog converters (DACs). Digital control signals can also be specified, to control things like switchers. HARPO also can sample the analog world and store what it finds there, for later use.

2. WHAT HARPO IS

HARPO is a type of computer program known as an INTERPRETER. It is an interpreter written in assembly language for 8080 microprocessor based computers, a type relatively cheap and availiable. The commands and scores which HARPO interprets are a more general language which is designed with video effects In mind. We need to define a few concepts and jargon terms essential to HARPO. A TICK is a sixtieth of a second (or whatever interval is fed into the master timing input in the computer). All times and rates in HARPO refer to TICKS. TICKS are synced up with the video equipment to allow HARPO to know where the video is. A POINT is a value, a number, a voltage if output to a DAC. A RAMP is a series of POINTs which is linear, i.e. if we graphed a RAMP with values of POINTs vs. time (TICKs) we would see a straight line. If we send a RAMP to a DAC we will get a (linearly) changing voltage. A WAVE is a more general series of POINTs and can be non-linear; that is, its graph need not be a straight line. While we can easily describe and compute the points in a RAMP (it's just a straight line), WAVEs are harder to describe. That's why analog synthesis techniques are so much easier from a user's standpoint than digital synthesis techniques. So we generally create WAVEs in the analog hardware and periodically sample the resultant voltages with HARPO. If we take the WAVE

Figure 1. Page 1 of the HARPO operating manual (1979). (courtesy. Carl Geiger).

CG: Well, yeah, until just recently all of our display systems have been analog. I mean we haven't had digital televisions for what, more than ten years or so.

MJ: But I guess what I'm saying is that in the beginning of video artists using computers – I guess the capability wasn't there to do more graphical work – it was more about using control voltages. So like you said, you could turn a knob faster, and then go back out to an analog device.

which HARPO saves, and send it to a DAC, we will get an analog approximation of the original series of voltages.

3. HOW USERS COMMUNICATE WITH HARPO

HARPO can operate in two modes. As an INTERACTIVE interpreter, HARPO accepts commands the user enters from a keyboard and executes the commands immediately. In SCORE mode a list of commands (a SCORE) is specifed ahead of time using another computer program called an EDITOR. The user then can tell HARPO to interpret the SCORE s/he has created. The SCORE can be saved and reexecuted later. Most command lines are interpreted identically in both modes, but in SCORE mode, command lines can be interpreted much faster than a user could type. This allows rapid changes in signals controling video hardware, which is what video effects usually consist of. In addition, the user can talk to HARPO via an analog device by using the MAKEWave command.
INWAVe commands.

4. HOW WE HAVE USED HARPO SO FAR

In our imaging system we use HARPO to control 13 DACs, which in turn direct various analog devices. We also use HARPO to control a switcher, and to provide precise timing for our effects. We usually work in this sequence:
We decide what we want our effect to look like, and what the time relationships are to be. Then we think about how we could produce these images using the video hardware we've got. Then if these pieces of hardware need to be controlled precisely, or quickly, or if there are more things to control than we have hands, or some combination of these, then we turn to HARPO to execute a series of commands to do these things. We can easily try out various pieces of the puzzle in interactive mode. Then we have to write a SCORE (a list of commands) with the EDITOR program, and make all the necessary patches between the computer and the video hardware. HARPO interprets the score, controls the video hardware, and we tape the results. Often while HARPO is controlling part of the hardware needed for the effects, we are controlling other devices manually, because many effects are more easily controled by hand than by machine. Typically our first attempts are lousy, and we have to re-edit the score (again using the EDITOR) and have HARPO try again with the revised score.

5. WHAT HARPO COMMANDS LOOK LIKE

Figure 2. Page 2 of the HARPO operating manual (1979).
(courtesy. Carl Geiger).

CG: Right. Initially… well, I guess digital in my system developed in parallel. I mean, you could do cool stuff with just digital on/off things into video and it could make interesting images just in pure digital. And that's digital circuitry going into analog. But the thing I just described – which is the computer language HARPO – there's no direct digital involvement with the image. All that the synthesizer sees is control voltages and in effect they're not coming from the keyboard or me turning a knob; they're coming from the computer. And so the real digital circuits don't have a direct impact on [the video]. They're just analog voltages that are controlling analog circuits in an accurate way.

MJ: So were you thinking of the synthesizer or the software being available to other people? Or was it mostly as an exploration as an individual artist for your own work?

CG: The computer language HARPO we totally designed for other artists to use. I mean, we were excited about the concept. As far as I knew nothing like that existed, and it could make a complicated thing of running this computer relatively simple for someone to do. I mean, there's actually that kind of stuff being done now in open-source software. There's performance video and performance software that anyone can use for free that is the same idea.

MJ: Like [the software] Jitter, for instance...When you say we, do you mean Synapse,[1] or...?

CG: Well, when I was developing HARPO, I could program but I wasn't that good at programming. So when I got the grant from the National Endowment for the Arts I went and hired a programmer. I worked with this guy, Rod Fountain. He was a very good programmer and he had worked in industry and he was done, was sick of industry, so he was psyched to have a project that was creative. So I would build the circuitry and he would write the software, and we'd go back and forth to make it all work. But the idea was ultimately, and when we wrote the grant: 'Hey look, we're trying to make a computer language for artists to do performances and control video and stuff in real time.'

MJ: So how was it used ultimately? How and where?

CG: I used it for my work and we would use it on occasion with some video artists who would come through Synapse. But the truth is not much was really done publicly. Computers evolved very quickly and, probably, I had to spend way more of my time making a living and I wasn't really able to... I mean the idea was to make it simple, but the truth is it wasn't really that simple. And it wasn't nearly as easy as our computers with graphical interfaces now where you don't have to know much about anything. You just have to move a few icons around – but in those days it was a simple idea but it wasn't as simple as it should have been.

References

Fountain, Rod (1979), 'HARPO Description and Operating Instructions', Lafayette, NY: Collection of Carl Geiger.

Geiger, Carl (2011), 'Interview with Carl Geiger by Mona Jimenez', February 5, Lafayette, NY: Experimental Television Center Archives.

Notes

1. Synapse was a media arts center that Carl Geiger and Lance Wisniewski started in 1970 at Syracuse University (SU) in Syracuse, NY, along with other SU students, including Gail Waldron and Bob Burns. Among the resources available to artists at Synapse in the 1970s were a broadcast studio, an in-house cable system and portable video equipment (Geiger 2011). Organizational records on Synapse, including videotapes, can be found in the SU Library Special Collections Research Center. Jitter is software that allows for real-time processing of video and is made by the company Cycling '74. See http://cycling74.com/products/max/video-jitter/.

Finding the Tiny Dot: Designing Pantomation

Mona Jimenez

In the late 1960s, Phil Edelstein and Tom DeWitt both found themselves in the Electronic Music Studio (EMS) at the State University of New York (SUNY) at Albany. The Coordinated Electronic Music Studio System (CEMS), an audio synthesizer built by Robert Moog to EMS director Joel Chadabe's specifications, hadn't been in place very long. State universities were expanding, and SUNY Albany had invested in a brand new state-of-the-art performing arts center. With state arts funding on the rise Chadabe was able to bring in big names in the avant-garde: Merce Cunningham, John Cage, David Tudor. Edelstein's interest in theatre, music and electronics was growing, and he was a 'gearhead' and a student of computer science.

When he arrived at SUNY Albany, DeWitt was in the middle of his film *LEAP*, which needed a soundtrack. He had apprenticed with the filmmaker Stan Vanderbeek, who knew Chadabe; this connection got DeWitt access to the CEMS. DeWitt had some cachet, as he had made several award-winning films in San Francisco and had been one of the filmmakers Gene Youngblood had written about in his book *Expanded Cinema*.

For Edelstein and DeWitt, their artistic practices and aesthetics could not have been more different. In Edelstein's view:

> [Tom] was looking for additional tools to tell stories. I was looking for additional tools to establish processes and establish places where people would interact in environments where the system, where the algorithm, would essentially identify the ways the participants – both the performers and the audience – were interacting with each other. (Edelstein 2012)

Edelstein understood electronic music, but he was introduced to electronic image making from DeWitt. Edelstein became intrigued when he saw DeWitt was feeding audio signals from the CEMS into an oscilloscope and recording the oscilloscope's image with a video camera. Edelstein and DeWitt began working on each other's projects and collaborating artistically, eventually working with George Kindler, Maude Baum, and others as the sound-image-dance group, Electronic Body Arts, also based in the Albany, New York area.

By 1973, DeWitt had a residency at the Television Laboratory, a new program at WNET, New York City's public television station, where artists were offered use of broadcast facilities for experimental projects. Nam June Paik had been the first artist-in-residence at the TV Lab and his new video instrument, the Paik/Abe Video Synthesizer, was part of the tool set available to visiting artists. DeWitt was working on *Philharmonia*

(1974), and he and Edelstein would make regular trips to New York City. The TV Lab had a PDP-8 computer, and WNET had received funds from the New York State Council on the Arts to interface it with the Paik/Abe (NYSCA 1973); in fact the TV Lab staff were trying to use it to automate a video switcher. Edelstein got hired by WNET to do the programming for the interface from the PDP-8 to the switcher, but ultimately the experiment failed. Edelstein and DeWitt ended up taking the PDP-8 back to Albany for repair, where they found a use for it:

> We realized we could start doing image capture. We were looking for better and better ways of doing man-machine interfaces. [...] It also became clear that we could take an image and derive useful information from an image for controlling an audio processer or patch or additional images. [...] You might put lines on a page and interpret how those lines go sonically. Tom came along and recognized that there was skyline notation, which is a series of stairsteps. He said 'We can build a reader that would allow a composer to set those stairsteps. [Then we could] point a camera at it and it would play that score'. (Edelstein 2012) [1]

The process of interpreting marks on paper as sounds involved keying and counting:

> You're taking a pitch over time. The simple engineering involved was that if you realize that [one area] was white and [one area] was black, [I] turn it on its side and I then count how many pulses there were before it went from white to black. And then I would find out if it was six pulses or 128 pulses, and we'd put out that number at that voltage. The farther the black went, the higher the voltage, the higher the pitch. And then I could take a piece of paper which had a series of stairsteps, put it in front of a camera and I would get different tone rows out based on those relationships. (Edelstein 2012)

From that process, a whole new set of ideas came to the forefront. The marks became points on a television screen: positions on an x-y axis.

> We knew that we could look at a point and find that point from a graph reader. We also knew that we could color encode things because we used chroma key at WNET. So it was just one of those things where you reverse the paradigm and instead of looking at the background and blocking it out, you look at a tiny little color foreground and extract that. [...] That had the convenience of lowering the demand on the dull, slow computer, the PDP-8. [...] We'd throw away everything except you grab this *tiny* little bit of relevant stuff. (Ditto 2012)

Anyone who has watched broadcast television has seen chroma keying. It allows the combination of two video sources; for example, it makes video appear in the box next to a television newscaster, or creates the illusion that the weather reporter is standing in front of a map. With conventional chroma key, the weather reporter stands in front of a screen of a particular blue hue. The blue screen is replaced by a second video source, the map. Edelstein and DeWitt isolated a very small dot as the key color, and they had the beginning of an input device that tracked motion. (See Color Plate 29.)

> Once you had that basic circuit running, then you could put a dancer, a human body in front of that, and you could detect one edge of the body. And then you could start capturing gesture. Now we're on to something. [...] Same basic circuit, but instead of some simple stairsteps I'm now going to point it at a body moving in space and I'm going to have this whole richer opportunity for woman-machine interaction. I'm winding up with a set of values defining their edge. [...] And that's just enough to start capturing gesture. And that was the beginnings of Pantomation. (Edelstein 2012)

With performers holding a tiny colored dot of a certain hue, the synthesizer could capture incremental points along the arc of a gesture, and superimpose theses traces over live video. Edelstein described the process broadly as 'deriving information from video images based on the position of a key signal relative to the video frame', which enables both 'generative and manipulative processes' (Edelstein n.d.).

In 1975, Edelstein and DeWitt wrote a proposal for the Design Device, a comprehensive system for image processing or a 'Graphic Guitar' (DeWitt 1975). The Design Device proposed to take what Edelstein and DeWitt thought were the best ideas in video instrument design and combine them into one. However, DeWitt's proposals to the National Endowment for the Arts and NYSCA did not deliver what they expected. They were forced to scale down the project and build only the motion-sensing part of the system, which became Pantomation.

Pantomation incorporated a Hearn Videolab (a synthesizer that included colorizer and chroma key circuits) with Serge audio modules, parts of a Rutt/Etra Video Synthesizer (a scan modulator) and newly designed components. The Pantomation team included Edelstein, DeWitt, Kindler, Richard Lanehart and Roger Myers:

> We had a little nugget of a team and then we had an *éminence grise* Joel [Chadabe], who let us do all these things, and kind of cheered us on and gave us this thought that we were kind of realizing in his own little, tiny little technical room what Bob Moog was doing in a great big factory. (Ditto 2012)

Pantomation debuted at the WNET TV Lab in a performance on New Year's Eve 1976 into New Year's Day 1977, with DeWitt appearing as the character Zierot le Fou in his new work *Outta Space*.[2] The PDP-8 went permanently back to Albany and Pantomation had its first live public presentation at SUNY Albany in September 1977.

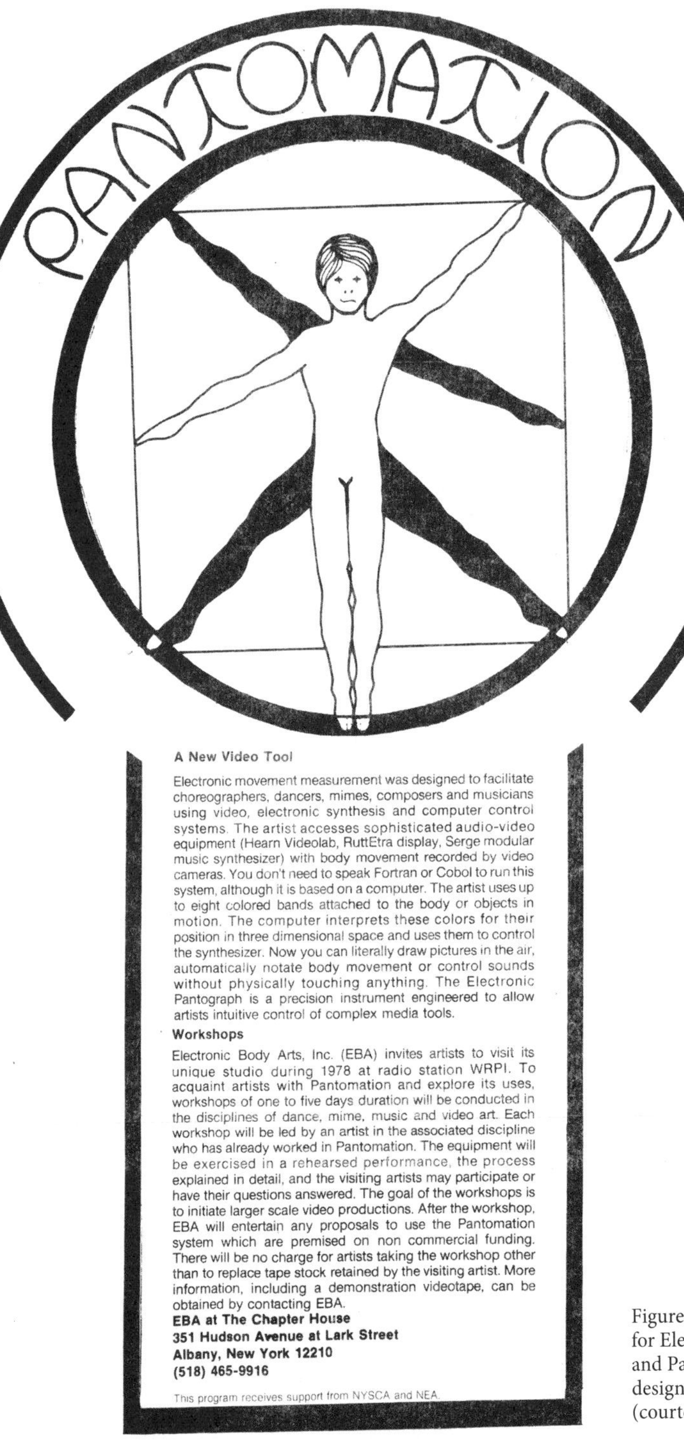

PANTOMATION

A New Video Tool

Electronic movement measurement was designed to facilitate choreographers, dancers, mimes, composers and musicians using video, electronic synthesis and computer control systems. The artist accesses sophisticated audio-video equipment (Hearn Videolab, RuttEtra display, Serge modular music synthesizer) with body movement recorded by video cameras. You don't need to speak Fortran or Cobol to run this system, although it is based on a computer. The artist uses up to eight colored bands attached to the body or objects in motion. The computer interprets these colors for their position in three dimensional space and uses them to control the synthesizer. Now you can literally draw pictures in the air, automatically notate body movement or control sounds without physically touching anything. The Electronic Pantograph is a precision instrument engineered to allow artists intuitive control of complex media tools.

Workshops

Electronic Body Arts, Inc. (EBA) invites artists to visit its unique studio during 1978 at radio station WRPI. To acquaint artists with Pantomation and explore its uses, workshops of one to five days duration will be conducted in the disciplines of dance, mime, music and video art. Each workshop will be led by an artist in the associated discipline who has already worked in Pantomation. The equipment will be exercised in a rehearsed performance, the process explained in detail, and the visiting artists may participate or have their questions answered. The goal of the workshops is to initiate larger scale video productions. After the workshop, EBA will entertain any proposals to use the Pantomation system which are premised on non commercial funding. There will be no charge for artists taking the workshop other than to replace tape stock retained by the visiting artist. More information, including a demonstration videotape, can be obtained by contacting EBA.

EBA at The Chapter House
351 Hudson Avenue at Lark Street
Albany, New York 12210
(518) 465-9916

This program receives support from NYSCA and NEA.

Figure 1. Informational material for Electronic Body Arts (EBA) and Pantomation, featuring the EBA logo designed by Vibeke Sorensen (c. 1970s). (courtesy. Experimental Television Center).

DeWitt later ported the system to an Apple II computer for better computing power and for ease of transport. DeWitt describes the 1980s version as a 'tracking chroma key system' that is 'similar to a light pen interface for a computer but uses a color video source and chromakeyer [sic] to detect positional information'. (DeWitt 1982: 61)[3]

DeWitt went on to make video works through various artistic collaborations, including as part of Electronic Body Arts and at Pantomation's new home in the Video Synthesis Laboratory at Rensselaer Polytechnic Institute in Troy, New York. DeWitt (now known as Tom Ditto) has continued with art making and with imaging inventions, most recently involving 3D spectroscopy. Edelstein went on to pursue computer programming, video, and electronic music composition and performance, including with the group Composers Inside Electronics. The moment of invention of the Pantomation is remembered by both Edelstein and Ditto as a time of intense experimentation and cross-pollination.

> Magic happened! And we were doing great stuff. No one of us was really responsible for any of this. A bunch of young people that had a lot of ambition to create things and no ambition to get rich. So that resulted in some really fine art-technology collaborations. (Ditto 2012)

References

DeWitt, Tom and Edelstein, Phil (1982), 'Pantomation: A System for Position Tracking', *Proceedings: 2nd Symposium on Small Computers in the Arts*, July, Los Angeles CA: IEEE Computer Society.

DeWitt, Tom (1975), 'Proposal for WMHT to the New York State Council on the Arts for a Unique Video Synthesizer', http://www.experimentaltvcenter.org/design-device-unique-video-synthesizer. Accessed March 20, 2013.

Ditto, Tom (2012), 'Interview with Tom Ditto by Mona Jimenez', July 23, Brooklyn NY and Ancramdale NY: Experimental Television Center Archives.

Edelstein, Phil (n.d.), 'The Key Positioner, a Device for Implementing Pantomation', http://www.experimentaltvcenter.org/pantograph. Accessed March 20, 2013.

Edelstein, Phil (2012), 'Interview with Phil Edelstein by Mona Jimenez', June 2, Port Washington, NY: Experimental Television Center Archives.

New York State Council on the Arts (NYSCA) (1974), *New York State Council on the Arts Annual Report 1973–74*, New York: NYSCA.

Notes

1. DeWitt explains his idea of using stairsteps in Tom DeWitt (1987), 'Visual Music: Searching for an Aesthetic', *Leonardo*, 20: 2, p. 116.
2. Dewitt gives a more full accounting of this night in Howard Weinberg's essay 'TV Lab: Image-making Tools' in this book.
3. The article (DeWitt and Edelstein 1982) has detailed explanations of the concepts and functions of Pantomation and includes block diagram and circuits. Additional information can be found on the Experimental Television Center's Video History Project website – http://www.experimentaltvcenter.org/history. Videos of Pantomation can be found on YouTube.

Preserving Machines

Mona Jimenez

> I think of the Rutt/Etra [Video Synthesizer] and Hearn [Videolab] combination as a guitar. [...] The sounding board was the Rutt/Etra and the strings and the neck were the analog synthesizers and the video synthesizers that drove them. And you could sit there and string your little wires together and come up with compelling abstract works of art. And I see no reason not to continue in that direction. And to teach people how that's done even though it's arcane. Well, let's face it; these wooden instruments are arcane too. [...] The piano got invented and it didn't mean they should throw away the clarinet. (Ditto 2012)

Conservation and restoration happen within the walls of museums and archives, with the creators and designers of objects, and within the garages and studios of enthusiasts and collectors. Each group or individual has its own networks, goals, philosophies and practices. There is no doubt tremendous variation in the way protective and restorative practices are carried out, depending on one's concerns, aesthetics, training, and orientation to professional bodies or communities of interest.

This essay attempts to cast light on the efforts of different communities of interest that are working to extend the life of electronic video instruments, or those who might be pressed into service. For the purpose of this essay, electronic video instruments are custom devices developed in the United States from the late 1960s to the mid-1980s that were used to make video art or other forms of time-based media or were used in performance. The devices include synthesizers, colorizers, keyers, sequencers, video-capture devices, computer interfaces and oscillators, to name a few.[1] They may be modified off-the-shelf devices or machines built from scratch. They were (or are) used in artists' studios, in media arts centers, in television laboratories, in educational settings, in arts exhibitions, for motion graphics and for live performance by artists and designers from many different disciplines. The instruments have been used to create thousands of time-based media art works by hundreds of artists. The art works are in museums and arts centers internationally, and are used widely in exhibition, education, and public programming, and for research.[2]

Electronic video instruments have been included in research initiatives such as DOCAM (Documentation and Conservation of Media Arts Heritage) based in Canada, and in exhibitions; for example, the 1992 exhibition 'Eigenwelt der Apparatewelt: Pioneers of Electronic Art' organized as part of the festival 'Ars Electronica' in Linz,

Austria. Some tools are still used in art schools in analog/digital labs, such as at the California Institute of the Arts, the San Francisco Art Institute, Alfred University (New York State), the University of Illinois at Chicago Circle Campus, and the Chicago Art Institute. The tools are part of the history of technology and the history of media art.

One may ask, why save art-making machines? Isn't the point the art work itself? By another token, one might ask why pianos or oboes are collected in instrument collections; why museums of technology collect Apple computers; why science museums collect printing presses; why design museums collect cameras; or why games are collected at the Museum of the Moving Image? Video instruments are image making and sonic devices, performance-enabling objects, evidence of electronic design, forms of industrial craft, and remnants of a do-it-yourself ethic. They are part of the balance against media histories that privilege the hegemony of broadcast television or those that trivialize the material and labor processes of art and media.

In 1992, Woody Vasulka accurately observed that in comparison to electronic audio instruments such as the Moog Modular Synthesizer, 'There's no comparable or intellectual protocol to even consider the video instruments as cultural artifacts' (Dunn 1992: 11). Twenty years later this may still be true, if we check the United States major institutions of art, science, technology, industry and history. The early video tools, like many forms of noncommercial

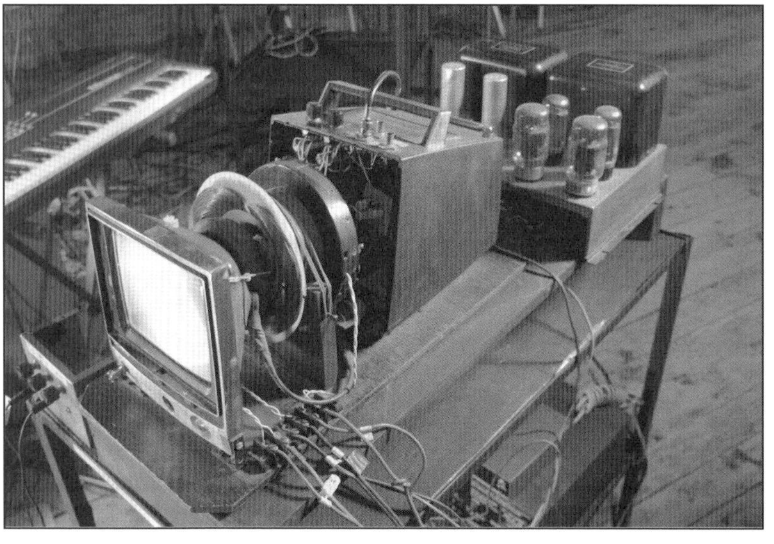

Figure 1. A Raster Manipulation Unit (aka the Wobbulator) in the studio of the Experimental Television Center (2007). (photo. Olivia Robinson; courtesy. Experimental Television Center). See Color Plates 35 and 36.

media technologies, do not fall neatly into a known category of history and have not been collected by memory institutions. They seem to fall through the cracks of the collecting missions of existing museums for television, computers and other related technologies.

Yet one must consider popular response. In 2010, when artist Blair Neal produced a short videotape about one of the very earliest video instruments, the Raster Manipulation Unit (popularly known as the Wobbulator) and posted it to the Internet, the video went viral (Neal 2010). The tape shows a 1960s black-and-white TV that has been broken apart, modified and hooked up to an amplifier and oscillators. Neal captured how the Wobbulator turns a familiar household consumer object into an image-generating device, and he found that his peers were fascinated.

A similar viral response happened when a friend of Tom Ditto's (the artist formerly known as Tom DeWitt) posted the following announcement in a technology forum: 'Kinect's Grandaddy Running On an Apple IIe In 1978'. The grandchild, the Kinect, is a phenomenally popular motion-sensing input device for PCs and the Xbox video game consoles made by Microsoft. The "grandaddy" is a video instrument called Pantomation that was designed with public arts funds by Ditto and Phil Edelstein – artists and technologists – in 1974 in the Electronic Music Studio of the State University of New York at Albany. The device, used by video artists and dancers, was a synthesizer that tracked and captured an object in motion, and combined the synthesized object with video in real time.

In addition to the attention garnered by older devices, designers such as Dave Jones[3] are still at work on new custom analog instruments. Thus, while some of us ponder the physical preservation of older machines, simultaneously the concepts and processes they represent are being incorporated into new analog devices. The processes that once involved numerous bulky devices are embodied in units that can be held in one hand.

In this essay, we will first look at conservation efforts under way among inventors, enthusiasts and caretakers, followed by a discussion of the practices and guiding principles of potential allies in conservation, and a discussion of new video instruments. Lastly, proposals will be made for a way forward. Although the essay talks in detail about conservation theory, only a few of the machines have been collected by cultural heritage institutions, and artists are still looking for real-time video tools. Thus, it will be proposed that the most appropriate site within which to advance preservation practices for the instruments will be one where art making, tool development, and conservation are carried out together.

A resurrection party

When I met Daniel Summer in September 2007, he was hunched over the open chassis of the power supply and amplifier unit of a Rutt/Etra Video Synthesizer. Matt Schlanger had opened up his company in New York City, Black Hammer, for a 'Rutt/Etra Restoration Party'. (Rutt/Etra is the synthesizer's affectionate nickname.) The synthesizer Daniel was

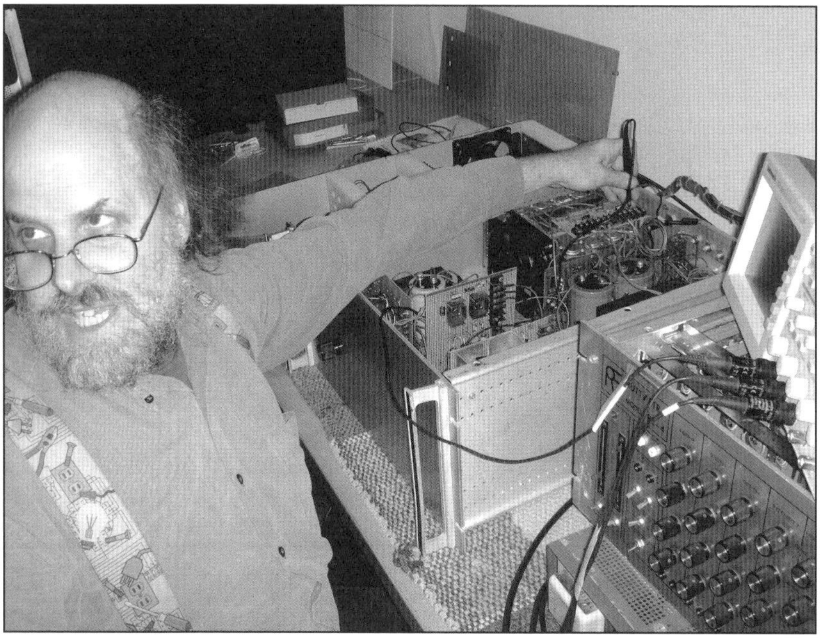

Figure 2. Jeffrey Schier trouble-shooting a Rutt-Etra Video Synthesizer at the Rutt/Etra Restoration Party (2007). (photo & copyright. Daniel Summer; courtesy. Daniel Summer).

working on had a WNET property tag on it – it was one of the first Rutt/Etras to be built in the early 1970s.

In 1972, New York's public television station, WNET, commissioned Steve Rutt of Rutt Electrophysics to build a synthesizer to add to the tool set for the Television Laboratory, an artist-in-residence program. The synthesizer would add a new tool to the conventional broadcast television studio that already held a Paik/Abe Video Synthesizer, a switcher and some video decks. Bill Etra had convinced the powers that be at WNET that he and Rutt could design a new, powerful machine to add to the arsenal. No doubt they had the enthusiastic support of the lab's artist-in-residence, Nam June Paik.

At the restoration party, Summer was swapping out electrolytic capacitors and checking the power supply. The changing of capacitors was preventive and being done before the machines were turned on. Summer handed me a bag of leaky capacitors and explained: 'When they go bang, it's sort of an ugly thing – they leak a chemical, and then there's a small explosion which can sometimes at a minimum make a big mess; at worst, knock out a bunch of components next to them' (Summer 2007a: 2).

A few feet away, Steve Rutt and Kyle Lapidus were troubleshooting a Rutt/Etra that had previously been owned by the musician Todd Rundgren. (Both this unit and the one with the WNET tag belonged to artist Gary Hill.) The Rundgren unit had been built in 1978 as a special deal, four years after Rutt Electrophysics had stopped producing the synthesizer; it had standard parts but also had been modified from the classic Rutt/Etra with some hand-wired, custom boards. Rutt was trying to figure out why one power supply was not getting warm and one was running hot. He was trying to avoid blowing out the high-resolution display, a rather pricey item. The old power supplies had been connected with wires covered with gaffer's tape that was rotting off.

Rutt hadn't been able to stay away from the party:

> I think I've never seen a room with this many Rutt/Etras. Because when we'd build them it would be a combination of somebody really anxious to get their hands on it, and we were always concerned about getting paid… we have, like, an entire farm full of them *here*. (Rutt 2007: 9)

There were two more in the room: a unit that was Bill Etra's that he had given to Summer, and a partial unit that Tom Ditto and Phil Edelstein had built as part of Pantomation.

At this point, the restoration party was about two weeks in. Summer had the idea for the gathering, spurred on by a need for Bill Etra to be able to have a working machine for a project that had come his way. Etra's machine had been in use by Summer and friends Benton Bainbridge and Owen Bush for much of the earlier decade. Rutt Electrophysics had built sixteen machines between 1972 and 1974, and the one in 1978 for Rundgren. Still, Summer couldn't find any other functioning machine for Etra to use and Summer's unit needed attention.

> The idea was to consolidate knowledge [about] the machines […] to extend the useful lives of the machines, and leave behind documents and knowledge which might increase the possibility of machines being serviced in the future after the original engineers were no longer around to help. Even with schematics available, there was a lot of variation between the machines so there was less reverse engineering to do having the designers present along with engineers who are currently active. The mix of talent and experience was pretty remarkable. (Summer 2012)

I returned on October 3, and they had another unit: Rutt's own machine had been dug out of storage. The group had made some progress with Rundgren's machine, most importantly because they had discovered that it had been modified by someone very well known in the network of designers and builders of 1970s video instruments – Dave Jones. Jones is renowned in these circles and beyond for his custom video tools such as the Jones Frame Buffer and the Jones Colorizer. (Jones also worked with Hill on custom hardware and software for many of Hill's video installations, so their connection did not come as a surprise.) Jones remembered a good deal of the details of the Rundgren machine and made extensive notes on his modifications for the restoration team's use.

That day the Rutt/Etra that Summer and friends had been working on was ready to fire up. Summer had been warned by Rutt to stand away from the machine, as the 10,000-volt power supply had the potential of arcing: 'Rutt assures me it can't go more than a few inches, but still I prefer to stay more than a few feet from it' (Summer 2007: 3). Summer was looking for some activity on the display, but as soon as he hit the power we heard a popping noise. He wasn't happy with the 'snap, crackle and pop'. Summer, a collector but not an engineer, decided he had reached the limit of his knowledge about old power supplies. He would change them out with some he found on eBay that he hoped were close enough to work. The old ones he would box up and send off to an expert to see if they could be repaired.

Summer said that the team had decided that installing contemporary power supplies was not an option:

> It's probably the case that modern power supply designs would be an improvement on these machines. But we've decided that rather than go that route, [we would go] in the direction of making [the Rutt/Etras] 'as good as they could have been given their design' [...]. I look at this as an instrument. And I know if this were a musical amplifier, the power supply would be one of the most important parts in the tonality of the amplifier and in this case, it's a video instrument. I think these [power supplies] actually play a role in the look that's made by the machine. (Summer 2007b: 5)

The 'Rutt/Etra Restoration Party' continued, lasting roughly five weeks, and by the end Summer's unit and both of Hill's machines were fully functional and calibrated. Rutt's machine, which was a prototype, 'went from totally nonfunctional, incomplete and infested with rats to functional' (Summer 2012). Artists Woody Vasulka and Ditto participated, and these efforts may lead to their machines eventually being restored. Documentation of the event, which included video, audio, photos and technical documentation, is being managed by Summer.

Individual restorations of other video instruments may also be under way. According to the website documenting KQED San Francisco's National Center for Experiments in Television (NCET), in 2008 Larry Templeton, Rick Davis and Bob Pacelli restored a Templeton Mixer to working order.[4] Templeton, a video engineer, invented the Mixer, which was a colorizer, keyer and mixer in one, for NCET in 1970. The report on the restoration is in the context of a site that collects stories from the former NCET community; it is unclear whether additional documentation exists.

Institutional efforts

In 2011, a major cultural institution took up a project with a smaller and more specialized team for restoration of a Paik/Abe Video Synthesizer. The synthesizer was coinvented by Nam June Paik and engineer Shuya Abe in 1969 or 1970, for installation in the New Television Workshop, a program of the public television station WGBH in Boston, MA.[5]

The synthesizer was in essence a seven-channel mixer and colorizer, producing layers of wild, saturated colors that are characteristic of Paik's work at the time.

At the Nam June Paik Art Center in Yongin, Korea, Shuya Abe, technical advisor Jung Sung Lee, media artist Kumlyun Lee, technical expert Ki Jun Lee, and archivist Sang Ae Park collaborated on the 'Abe Video Synthesizer Restoration Project'. The restoration team identified three early models of the Paik/Abe Video Synthesizer and seven that are considered compact models. Five of the compact models were built by students at the California Institute for the Arts, and one of the students, Sharon Grace, made her machine available for the project.[6]

Although the Paik/Abe Video Synthesizer is commonly known in the United States as a stand-alone device, the restoration involved a whole set of related devices that were organized into several racks. These devices included a color encoder, a scan converter, a function generator, an audio signal generator, an audio amplifier, an audio mixer, and a video switcher. The goal of the project was achieved: to restore the compact version to functionality based on the original design. Taking place over a six-month period, the project culminated in a workshop where the audience had hands-on access to the controls, to 'provide the digital generation with opportunities to experience an early analog video synthesizing' (Park and Lee 2011). The resulting catalog, *Paik-Abe Video Synthesizer: As freely as Picasso, as colorfully as Renoir*, is full of photos and diagrams, suggesting that documentation is plentiful.[7]

In a much earlier effort, for the 1992 exhibition 'Eigenwelt der Apparatewelt: Pioneers of Electronic Art' numerous custom video tools developed in the United States in the 1970s and 1980s were gathered from disparate sources by curators Woody Vasulka and Steina Vasulka and their team. Nearly twenty machines were represented in the exhibition and many were installed and brought into service for the occasion.

Given the delicacy of the instruments, external interfaces were built for the working machines that enabled viewers to have hands-on access with video synthesis. In the exhibition catalog, several of the instruments are wishfully listed as 'awaiting restoration'. Documentation of the process for 'Eigenwelt der Apparatewelt', including technical documentation, is part of the Steina and Woody Vasulka fonds at the Cinémathèque québécoise in Montreal, Canada. In addition, many, if not all, of the documents can be found on the Vasulka's personal website.[8]

The Experimental Television Center (ETC) recently released an eight-disc DVD set, *Early Media Instruments,* that allows viewers to see eight custom video instruments in operation. The DVDs show a technical expert demonstrating the tool with the image output of the device superimposed on-screen. The videos are a very valuable way to gain information about the capabilities of each machine, its key components, and how the tool was integrated with other devices. As part of the production of the DVDs, ETC restored a number of devices to working order.

Also, as part of the ETC Residency program, many examples of video instruments were maintained, repaired and restored over the years. Director Ralph Hocking and associate director Sherry Miller Hocking have close ties with former staffers, volunteers

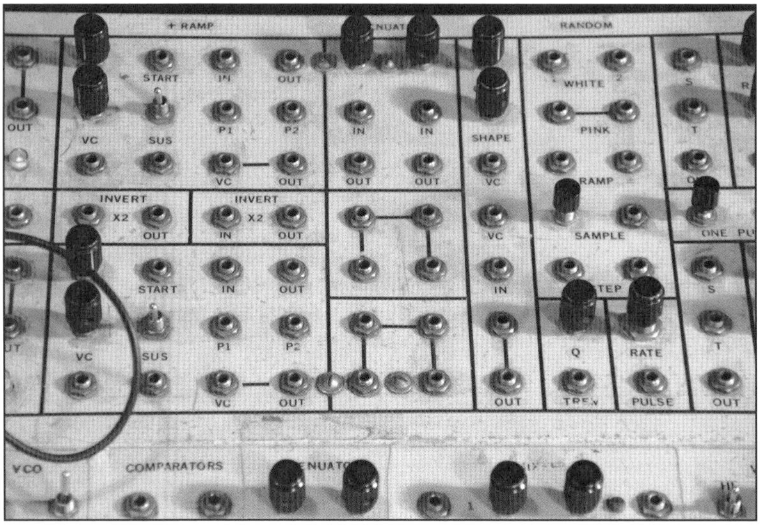

Figure 3. The top panel of an Analog Control Box that was built by Richard Brewster in 1977–78 for the ETC artists' studio. The boxes generated waveforms and other audio signals that were used to alter video images and were in use until the ETC studio closed in 2011. (photo. Olivia Robinson; courtesy. Experimental Television Center).

and visiting artists who hold knowledge about custom video tools. Miller Hocking has been meticulous about collecting documentation and sharing it through the 'Video History Project' website, and many important and revealing documents have been deposited at the Cinémathèque québécoise.[9]

All of the efforts described above are very encouraging and need to be continued. They come from the same impulse and passion – to keep the video tools accessible and to ensure that they become part of the historical record – even if caretakers follow different approaches, have different aims, and choose different methods of documentation and accessibility to that documentation. What is most important is action. However, thinking carefully about the needs of the machines, what's important to conserve, and the quality of our decision-making is also important. Documentation of the machines and any conservation and restoration actions must be captured and shared widely if the instruments are to be maintained and used in the short term, and in the long term to be understood by future researchers, educators and artists.

As noted above, these custom tools have not been collected by cultural heritage organizations, and whether they should be at this juncture is subject to debate. The likelihood is very slim that artists will have access to these tools to make new work once they are officially deemed legacy technology and historical artifacts. However, without institutional homes, it is also very unlikely that the machines will become accessible to researchers, educators, students and curators as part of the history of media art and the history of technology.

The troubles of video instruments

As noted above, machines don't take well to storage and disuse. Without the warmth of electricity and the movement it generates, mechanical parts stick in place; lubricants, meant to ease friction, become sludge. Once pliable, rubber rollers and belts dry and crack. Connectors oxidize and capacitors reach the end of their useful life, threatening to explode when the machine is fired up. Batteries begin to ooze and contaminate. Software becomes inaccessible when the hardware or software it depends on becomes unusable, and obsolescence makes portability to a newer computer system unfeasible. Whether the machine is well documented or scantily described (as is more often the case with custom-designed tools) shifts in electronics design and materials make maintenance, repair and reconstruction difficult or even impossible.

The video instruments that are the subject of this essay are largely out of use. Inventors, collectors and artists hold the machines in varying states of wholeness and usability in their closets, barns and studios. Stories like Dan Summer's above from the 'Rutt/Etra Restoration Party' remind us of the issues facing these devices: physical deterioration, the lack of parts, the incongruities between basic components from the past and now, and how a simple decision about the swapping out of a power supply may affect the 'tonality' of the instrument.

Until 2011, the Experimental Television Center (ETC) maintained a residency program and a hybrid analog/digital studio where an artist could generate new video and sound works by patching together legacy tools – such as 1970s colorizers, keyers and capture devices – with newer computer-based hardware and software. The studio offered both custom machines and off-the-shelf devices, and spanned tool development during ETC's 40-year history with media art. At the writing of this essay, the studio is dismantled and the machines are packed up and in storage; their disposition is unclear. Following the studio closure, ETC director Ralph Hocking put out a call for proposals for use of the devices, soliciting an entity or person that has the capacity to keep them functioning and available for creative use. In other words, to keep them functioning as 'living machines'. Hocking's call underlines the fact that reconstructing an electronic studio from stored equipment is not for the faint hearted:

> Be aware that very few of the machines are immediately usable. Most of them need connections to synchronization signals, signal routing and display. Some need time base corrected input signals, it's not just plug the camera in. Audio machines might need amplifiers and speakers at least. These machines need someone capable of defining and maintaining their purpose. Repair parts might not be readily available. (Hocking 2011)

Hocking's warnings allude to the knowledge needed to maintain older devices, and to integrate them with more contemporary machines in the service of creative expression. Mixed in with his cautions about the dependencies of the machines – that they need external devices for synchronization signals or for stabilization and display – is his pointed comment about the necessity of understanding a tool's function, limitations and possibilities; the tools need 'someone capable of defining and maintaining their purpose'. Simultaneous attention is needed not only to meet physical requirements and care, but also to maintain a tool's purpose and potential.

While each custom device in the ETC studio has its own set of functions, most were intentionally designed as general-purpose machines; that is, to have multiple uses and applications as determined by the artist. Many are physically programmable, where the artist's physical patching between a device's modules serves as aesthetic or conceptual instructions. For example, in a simple set up on the Sandin Image Processor, two separate video images could be combined by patching signals from two different cameras into an adder/multiplier module, thus layering or 'keying' the two images by, say, substituting all of the blackest parts of the first image with the second image. By patching the output of that module into another module, the composite image could be made negative. By patching in a waveform from a third module, an oscillator, the artist could automate a switch from a negative to positive and determine the rate of the switching.

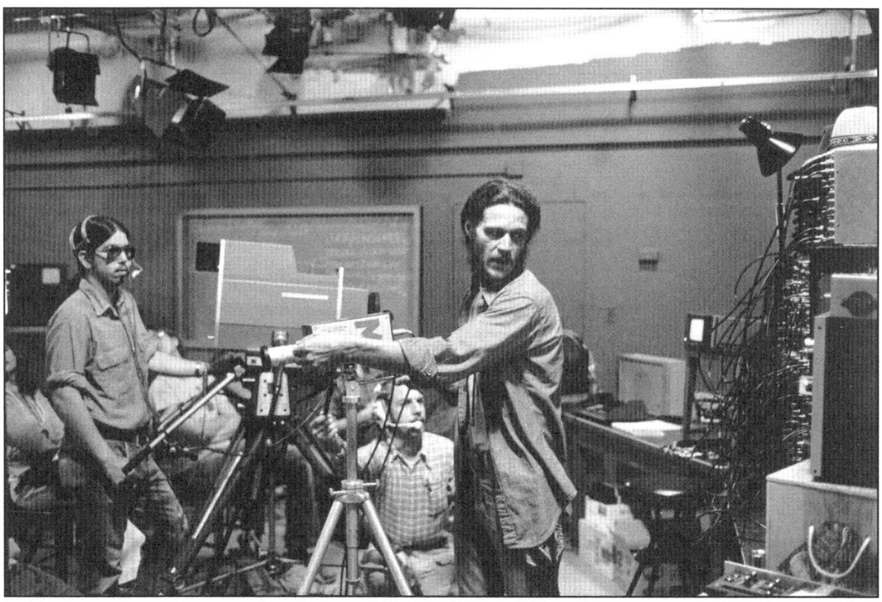

Figure 4. Phil Morton (standing right) adjusting a camera as an image source for the Sandin Image Processor at the Sacramento Electronic Visualization Workshop, California State University, led by him and Jane Veeder (1978). The unidentified men are documenting the workshop. (photo & courtesy. Jane Veeder).

In ETC's studio, the integrated tool set enabled a much broader form of programming, in which the output of any one device could be patched to any another device through selections on a matrix. Therefore, the artist chose and interpreted the functions of individual devices and of the system, both to achieve predetermined effects and to freely explore images and sound through the interplay among the various component parts. Thus, replicating ETC's studio would require knowledge not only of how to set up signal flow among multiple devices, but also the requirements of each device that would allow its full functionality to be available to the user. The idea of a flexible system-as-tool set was used in many different settings, from NCET to the Circle Graphics Habitat (University of Illinois at Chicago Circle Campus) or media arts centers like the Electron Movers in Providence, Rhode Island.

A distinguishing feature of these early analog devices is that the artist can work in real time – turning knobs, making switches, setting up sequences of images – and can see, hear and respond immediately. (Affecting digital video in real time has only recently become possible as a result of faster computer processors, and has been made easy by software such as Cycling '74's program, Jitter.[10]) In this way, the tools function as instruments. A musician hears a note and makes tonal adjustments on a musical instrument; an artist using electronic tools can see, for a simple example, the impact on the tint of an image of turning a knob on a colorizer. The artist can then turn the knob to shift the hue. Also, some devices are generative, whereby a set of instructions sets into motion a process where images are created over time. These generative and real-time design characteristics, along with the possibility to have one device impact and/or control another, gave the artist options for performativity that could not be found in off-the-shelf image-making devices of the 1960s to 1980s.

It remains to be seen if the living lab that was the Experimental Television Center studio will be reconstituted. If not, physical deterioration and the disappearance of those with knowledge of the tools' intricacies will take its toll. Machines that were working when the system was packed up in 2011 will become part of the category 'awaiting restoration'. As in the situation with early media art works themselves, which have been preserved in great part through advocacy and by bringing the knowledge of the media arts community to bear upon the larger media preservation field, the custodial role that collectors and artists assume for these instruments will make the difference between future use and obscurity.

Relevant areas of conservation practice

There is no single conservation practice that neatly encompasses the issues present with custom video instruments. However, we will see that there are potential overlaps with conservators of time-based media art and more conventional instruments; in addition, there is potential shared interest with conservators of science and technology. In each of these disciplines, caretakers have begun to flesh out key issues, including what risks and values there are in maintaining working objects.

Given the instrumentality and performability of the custom machines, conventional instrument conservation is one place to look for guidance. All instruments have an aesthetic dimension as objects, but they also have an experiential dimension. One may admire a violin for its graceful construction, but it is through the laying on of hands that one experiences its purpose. Custom video instruments have a physical presence and can be quite appealing as objects, but without the artist's choices remain undefined and dormant.

In addition to the aesthetics of the object, conventional instruments have a sort of qualitative ideal, a 'sound' that is brought out through the action of playing. Each video instrument also has its unique visual character as a generator of images and sounds, and can be thought of as having a certain range of potentiality that is enacted by the artist. For example, just as conventional musical instruments of different eras have a different sound, images created through the 1970s' Paik/Abe Video Synthesizer have a different visual quality than those made with a Jones Colorizer from the 1980s, although they are both custom devices that allow for color shifts.

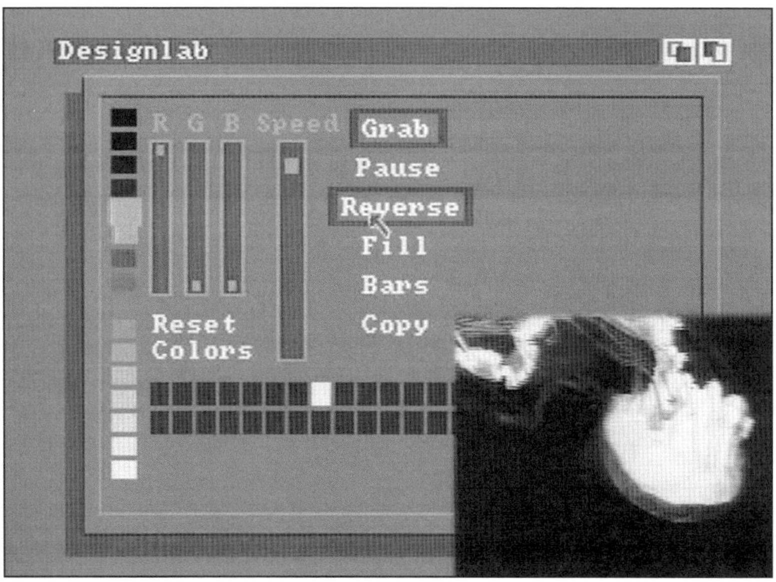

Figure 5. Still from *Early Media Instruments*. 'Jones Frame Buffer' (2010). Pictured is the Amiga computer-based software used to control the buffer, an image capture and layering device designed by Dave Jones (c. 1980s). (courtesy. Experimental Television Center).

Custom video tools also have commonality with (and in fact could be considered part of) the artifacts of industry and technology, which are the purview of industrial conservation.[11] Whether a tractor, a printing press or a video device, industrial and technological artifacts have both a dimension of craft and a dimension of process. The design and structure of the object can communicate much about an artifact's place in technological or industrial history, but without operation, the breadth of the artifact is diminished. One can admire a machine's external design but miss its internal distinctiveness. For example, the Jones Frame Buffer is an elegantly designed red box that is a product of craft, and the name 'frame buffer' evokes one of its principal features – the capture of individual video frames – but by no means describes its full potential. It is through the 'playing' of a working unit that the underlying processes of image layering that are unique to the tool are revealed. Furthermore, its operation is dependent on custom software developed by Jones for the Amiga computer platform; the software also has unique design features (see Figure 5).

Conservators of contemporary art have also been concerned with machines and equipment as components of time-based media art works. Often, to maintain a media art work's integrity, underlying technologies must be preserved that are critical to the work's functioning or to its meaning.[12] In the literature for time-based media art conservation, however, research has largely focused on the most common off-the-shelf devices: video playback decks, projectors, monitors, and the like.[13]

What keeps a video instrument alive and mysterious, versus codified and narrowly defined, is also a shared concern; what constitutes 'aliveness' has been an ongoing theme in time-based media art conservation. For example, analogies are common between media art installation and performance, where the ideas and processes of the work, rather than any particular manifestations, may be considered the essence of the work. If the artist is not available, the conservator might rely too rigidly upon previous components and insist upon certain parameters that deaden the piece.

Unfortunately, there is little literature to draw upon about conservation that directly addresses maintaining technological devices. One of the few articles concerns the ENIAC computer, completed in 1946 and considered the first electronic computer in the United States. Owned by the Smithsonian, the ENIAC was the object of conservation in the early 1990s, yet the emphasis for its conservation treatment was only on the external characteristics of the machine. (Understandably, bringing the ENIAC to working order would be a monumental task.) Treatments focused on restoring the 'in-use appearance' by replacing plugs and cables that visually matched those that would have been present if the unit was functional, such as giving careful attention to the vacuum tubes that were such a predominant part of the computer (Ecklund and Richwine 2004: 255).

Developing standards and practices for industrial conservation hold promise, and the warning bells sounded by this community sound very familiar. Speaking in 1989 to the loss of engineers and technicians who understand older industrial and technological machines, Jonathan Minns of the British Engineerium predicted a future of 'the blind leading the blinkered' (Minns 1989: 55). At the conference 'Industrial Collections: Care and Conservation', Sir Neal Cossons alluded to the meaning of machines beyond their technological story: 'Their messages and metaphors are as yet unexplored and they lie much deeper in the national subconscious than a simplistic taxonomy of machines might at first sight lead us to believe' (Cossons 1999: 9). Cossons called for 'a rational system of support and management'. Bénédicte Roland-Villemont and Claude Forrières, speaking at the same conference, outlined an interdisciplinary approach that incorporated various specialists and included assessment, treatments and documentation.

These themes – the threat of loss of technical knowledge, the need for standards and practices, and the value of an interdisciplinary approach – are reminiscent of discussions that have been occurring for decades among media preservationists, who rely upon obsolete playback equipment to transfer older tapes and maintain technology-dependent installations. Cross-departmental working groups in contemporary art museums focused on time-based media have sprung up and are dedicated to interdisciplinary approaches.

A theme that is central to the conservation of video instruments, and one that is common to both industrial conservation and conventional instrument conservation, is whether, when, and under what conditions an object should be operated (referred to as 'worked' for industrial conservation) or played (in the case of conventional instruments). In other words, are there dangers to keeping video instruments in working studios, if one is truly interested in having future generations understand the processes of twentieth century media artists? Or by not 'working' the instruments do we reduce their meaning and complexities and make them less understood and thus even more subject to loss?

To work or not to work

Jay Scott Odell and Cary Karp define instruments as functional objects, and place object function and design on an even par with aesthetics. In fact, for Odell and Karp the three characteristics cannot be separated. They point out that instruments 'require physical interaction to fulfill the purposes for which they were made' (Odell and Karp 1997: 1). Musical instruments and electronic-imaging instruments share these attributes of function, design and aesthetics.

Odell and Karp stress that use (and any restoration necessary to enable use) is risky and must not eliminate options for interpretation by future researchers and the public. In other words, would the restoration erase the cultural and material context in which the instrument was made? One could argue the reverse: that not allowing use would close off avenues of interpretation. Both types of instruments produce aesthetic experiences that are reliant upon interaction and both are silent and may be experienced as mundane without use. (While conventional instruments are often seen to have an aesthetic dimension, this may not be the case with video instruments. Although the devices are in some cases highly designed, others may appear to be without an aesthetic dimension – simply metal boxes with generic inputs and outputs, recycled parts and jumbles of cables.)

However, with instruments, restoration may need to precede use or even valuation. Odell and Karp caution that decision making should not be taken lightly:

> Restoring an instrument to functioning condition should not be considered unless an extremely important historic, technical or aesthetic quality can only be determined by actually operating the artifact, and only if this information cannot be gained in some other manner (Odell and Karp 1997: 4)

Odell and Karp also set out criteria that conservators can use to determine whether an instrument should be restored to functionality: the instrument's uniqueness or ubiquity, its ephemeral features and condition, the expected impact of use, the information to be gained through use, and whether expertise exists for restoration and maintenance.

Industrial conservator Richard Gibbon asserts that working is necessary for an object to reveal 'its function or the principles under which it operates' (Gibbon 1999: 18). In *Larger and Working Objects: A Guide to Their Preservation and Care* (1997), Stephen Ball, a conservator at the Royal Scottish Museum, wrote what appears to be the first written standards (in English) on the conservation of machines. Ball recognizes that as soon as an object is brought into operation, there are changes to its physical state – and these changes can occur both before and after the object's becoming part of a personal or institutional collection. In Ball's view, if a machine stops working or is retired:

> To a greater or lesser extent, the 'original' object is gone; what remains is a continuity of operation, usage and working practice. In other words, the essence of the working object is what it *does*. According to this argument, the body that operates the object is following good conservation practice, but it is conserving the form and function of the object rather than its physical make-up. (Ball 1997: 26)

Ball implies that simply restoring a machine to use still leaves much critical information hidden and at risk of loss; the machine has designs and processes that are not immediately seen and understood. In other words, by seeing a custom video instrument in action we can understand certain parts of the design, but there is much to know about its invention as expressed in unseen processes. Also, custom video tools reflect a certain period of circuit design, where hand-drawn and hand-taped circuits were the norm, a process now replaced by later computer-aided design.

On the subject of underlying processes, with a conventional instrument such as a violin, the full scope of its potential – what the instruments is capable of producing sonically – is only revealed by someone with deep knowledge of string instruments. With custom tools, this is also true; a machine can be turned on, but only its most obvious capabilities may be seen. For example, a colorizer, with its labeled knobs for hue and chroma, can be

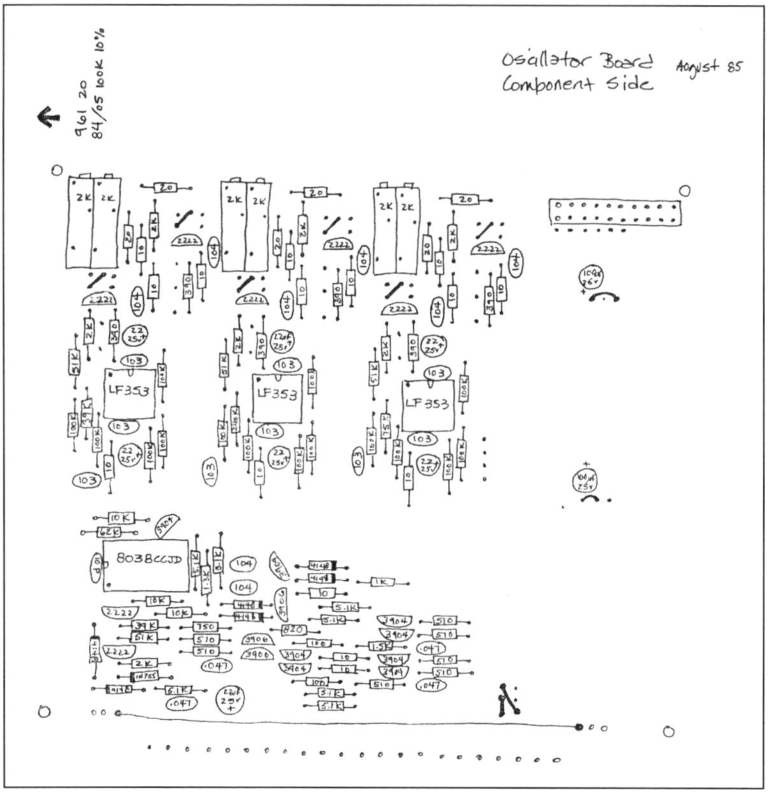

Figure 6. A drawing by Connie Coleman of a board in the Jones Oscillator Bank designed by Dave Jones for the Experimental Television Center (1985). (courtesy. Experimental Television Center).

assumed to output color images – an obvious function. However, an essential function of the Jones Colorizer, part of the tool set in the artists' studio at ETC – is its capability to automate and repeat changes in hue and other visual parameters through the application external waveforms. Without an oscillator to feed waveforms to the Jones Colorizer, and an artist aware of the potentiality, the tool is reduced from an unpredictable instrument (one that has open possibilities and gives the artist many choices) to a very predictable effects generator (a device that simply colorizes).

For Ball, human and financial resources are other factors that will determine whether an industrial artifact can be worked. While he states that the safest route (especially for objects that are one of a kind) is not to operate, Ball notes that the museum may need to choose between 'object conservation' and 'operation conservation'. Ball differentiates between works acquired for the long term and those whose purpose is 'research, public enjoyment or education', which he thinks may be more appropriate for operation conservation (Ball 1997: 26). Video instruments could easily be seen as objects of research or education, whether used by artists or the public.

Ball, Odell and Karp all assert that if an object is unique, there is a strong argument against functional restoration and use. The authors see risk both through use and through the risk of damage (Odell and Karp) and loss (Ball) during restoration and maintenance. With the 'Eigenwelt der Apparatewelt' exhibition, that's precisely why external interfaces were used. If one strictly followed the recommendations of these conservators, however, custom video instruments would never be restored or used, as most of the tools were produced in very small quantities and are scattered among numerous collections.

In a subsequent article, Mimi S. Waitzman and others lay out an ideal set of criteria for deciding whether an instrument should be worked, taking into consideration that:

- the instrument is already in working condition and its continuing use will not detract from its technical and cultural value;
- the instrument is in stable and safe condition;
- qualified expertise is available for regular tuning, maintenance, consultation and monitoring;
- the owner or custodian has a clearly articulated policy on the instrument's use, which meets the standards of ethics and practice of an appropriate national conservation or museum organization;
- the instrument is not to be treated as a working tool in the everyday world, but its historic value is understood and appreciated; and
- every effort is made to make players and audiences aware of the speculative nature of the instrument's 'original sound' [...]. (Waitzman et al. 1997: 83)

These are all common-sense instructions. Guidelines for instrument and industrial conservation, and for that matter the conservation of machines and equipment associated with time-based media work, are not substantially different from each other. In all three areas of conservation, an audit or assessment of a machine guides a care and treatment plan, and specifies how, when and by whom operation can occur. Also, each conservation discipline has standards of documentation. Ball is particularly thorough on this, calling for an operating manual that specifies the 'procedures, requirements, and conditions' under which it is to be operated, and noting that every time an object is worked, a record should be made (Ball 1997: 6). For video instruments, for example for the Paik/Abe Video Synthesizers at the Nam June Paik Center, a manual could guide the staff to the care of the machines by tracking when a machine is used in exhibition and for education.

Larger and Working Objects also gives very detailed guidelines for documenting maintenance, repairs and the working of machines. Because Ball believes that loss occurs most dramatically at the point of restoration, most of his recommendations concern the documentation of the object's assembly and disassembly, which he recommends be done only when no other options are present (Ball 1997: 15). It should be noted that Ball's colleague in industrial conservation, Richard Gibbon, sees restoration as a point not just of risk but of opportunity: 'the restoration process reveals undiscovered knowledge about the object' (Gibbon 1999: 23). This knowledge must be captured in a form that can be shared with not just other custodians and conservators, but with researchers as well. We saw this earlier when the 'Rutt/Etra Restoration Party' understood that Dave Jones had done the substantial modifications; it opened up a whole area of information about the machine that would not have been known otherwise – information essential to its conservation.

Ball makes detailed recommendations, covering such topics as an operating log ('a document recording the details of every occasion on which an object is worked') and an operating manual ('a document setting out the procedures, requirements and conditions attendant upon the operation of a working object') (Ball 1997: 6). He goes into great detail about the contents of each of these documents, down to saving removed parts; he instructs: 'record in detail what was found, what was removed, how this was done and with what it was replaced' (Ball 1997: 25). He also makes a checklist of condition assessment, noting such problems as corrosion, rot or mold, cracks or splits from temperature and humidity changes, and stress faults.

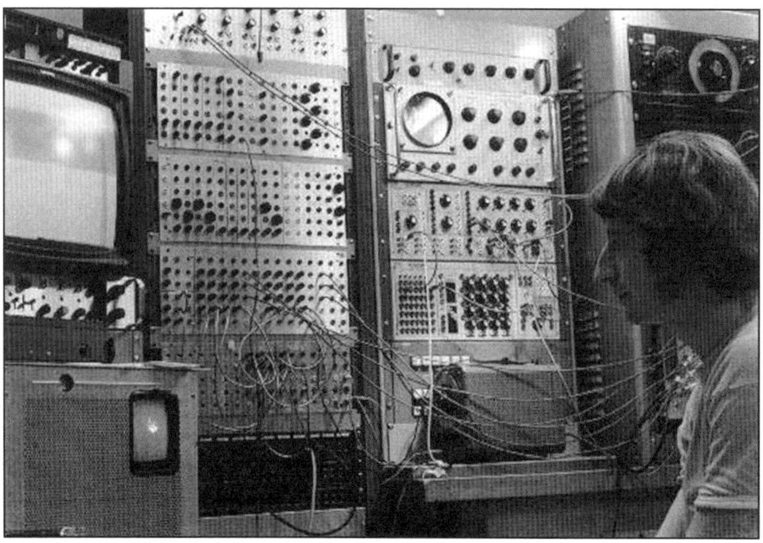

Figure 7. Tom DeWitt at the WRPI Video Synthesis Laboratory in front of Pantomation, a synthesizer system for motion capture that included a Rutt/Etra synthesizer and a Hearn Videolab (c. 1979). (photo. Vibeke Sorensen. courtesy. Experimental Television Center).

Guidelines for use of musical instruments are much more developed, although at this point they may have little relevance for video instruments. The International Committee for Musical Instrument Collections (CIMCIM) of the ICOM (the International Council of Museums) produced *Recommendations for Regulating the Access to Musical Instruments in Public Collections: 1985* that addresses the issue of instrument use.

With conservation there is no static, ideal object that is recreated through restoration. As mentioned above, time-based media art conservation is concerned with works of art, not art-making devices; therefore existing guidelines do not fit neatly for custom video tools. However, one could apply decision-making models usually reserved for media art to these devices. For example, drawing off of recent literature, Pip Laurenson has proposed a risk-assessment process whereby one looks at the anatomy of an art work, considering its key components and their relative value, in addition to any loss already sustained by the work and whether it can be reversed or mitigated. Considered as part of key components are its qualities, behaviors, look and feel, and interactivity (Laurenson 2007).

The key components and qualities in art works are analogous to key processes and characteristics of a given video instrument. If one analyzes a custom tool in the way Laurenson suggests, it may help to sort out the layers of meaning and value that are implicit in the conservation approaches above. Such an analysis would also be very useful for devices that have component hardware and software necessary for the tools' functioning, or where the 'feel' is part of the experience.

Designer Dave Jones remembers and associates the tactility of analog tools as a key component of the early instruments:

> People will make a digital device that simulates the way certain things looked in general, or in some cases, their interpretation of the way something looked. [...]They might say, 'Well that thing made it red, green and orange, so here is this software that makes it red, green and orange.' But that may not be what the original machine could really do. Plus a lot of the digital equipment doesn't have that physical feeling of a bank of knobs, or the real-time interaction of turning a knob and watching what's happening. Even the digital knobs available through MIDI and stuff like that, they don't have the same level of sensitivity as the real analog knobs. Even though you feel like you're turning something, you may be turning something with a couple hundred or maybe a thousand steps to it, whereas the original knob might have tens of thousands of steps because it was smooth analog. (Jones 2005)

Taken together, the conservation practices above suggest that the video instruments can be evaluated quite differently considering different institutional contexts. In an arts and culture context, the machines may be seen as key to understanding the early days of video art, and as sharing attributes with media art works. In a science and technology museum, the same tools may represent a certain period in electronic circuit design or in the development of imaging devices. In the context of an instrument collection, they may be seen as an extension of the development of electronic audio instruments. The point is that although these tools grew out of the media arts, they may eventually find their home in any number of institutions. Each institution will have a particular orientation toward the tools, how they should be conserved, whether they should be used, and what form of documentation is adequate to their understanding and interpretation.

Tools and emulators

As noted in the introduction, analog tool development has not completely stopped. Among those making analog tools is Dave Jones – designer of the Jones Colorizer, Jones Buffer and other media arts tools – who has just released an image-making device, the MVIP or Mini Video Image Processor. (See Color Plate 32.) The MVIP will fit in the palm of one's hand and replaces several of the devices in the ETC Studio. Jones intends to recreate all of the custom devices in the old ETC studio with palm-sized devices to complete a Modular Video Synth series.

Modules from the Sandin Image Processor are also being recreated in miniature form by Lars Larsen and Ed Leckie of LZX Industries,[14] based on Dan Sandin's original designs (see Figure 8). Both the MVIP and the LZX modules retain analog attributes on input and output and have the knob interface typical of earlier video instruments.

These efforts are exciting because they represent a continuation of the idea of open-ended imaging systems such as those in the ETC artists' studio. These efforts do not need to be seen as an alternative to a living lab on the model of ETC, or as a reason not to pursue conservation of the early video instruments. In fact, if the ETC residency program were still operating, the MVIP would be installed in a rack next to audio tools built in the early 1970s, as it was when ETC closed. In the spirit of maintaining old and new, analog and digital, at ETC, there's no reason to close off one strategy for maintaining media history in favor of another.

Jones distinguishes his analog device from digital hardware or software that imitates analog equipment or processes; with the MVIP, Jones has made a new analog instrument. Emulation has a visually different quality for Jones:

> [I]n some cases, emulation is a little too perfect. Part of the reality of the tools was those imperfections in the machines and in the video itself, little distortions that happen to video that created an additional look and feel to the image. If you emulate what the device was intended to do, you get the majority of the effect but you don't get those subtle things that happen, which weren't necessarily intentional but that came along with the fact that you built it and it was in an analog world [...] maybe there were color artifacts or little distortions to the image, like changes in the linearity of the

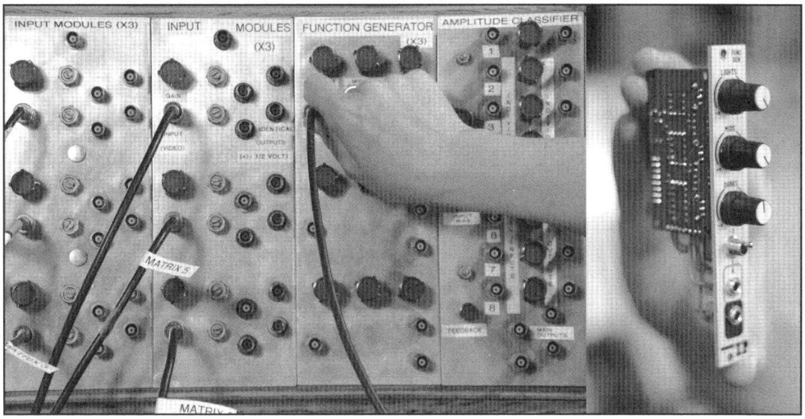

Figure 8. (l) Original modules from a Sandin Image Processor (IP) built by Dick Sippel (photo. Olivia Robinson. courtesy. Experimental Television Center); (r) a contemporary module for a function generator, one of the IP modules, designed by Lars Larsen and Ed Leckie based on original designs by Dan Sandin (2013). (photo. Lars Larsen; courtesy. LZX Industries).

> grey scale – noise, which gave character to the image processing. In some cases, the emulation is too smooth. (Jones 2005)

Here again, what one considers success hinges upon the key qualities that one associates with a particular instrument. Those key qualities are tied to certain processes within the instrument, and dependent on certain material conditions. Jones predicts that the emulations could eventually incorporate noise and aberrations, bringing their effects closer to the original devices. One would have to have deep knowledge of the qualities to be emulated:

> As emulation and the digital tools get better and better, you can probably start emulating the distortions and problems, and get even closer to what some of the original results were. This would be going beyond what the concept of the original machine was by trying to actually mimic the reality of the original machine. But at this point, that is still in the future. It's hard enough to even emulate the basic concepts of what the machine was trying to do, let alone emulate the reality of the small artifacts it had. (Jones 2005)

Emulation can mean much different things to different people. Bill Etra, the co-inventor of the Rutt/Etra Video Synthesizer, has released a software emulator of the 1970 hardware that was developed by him and Anton Marini, better known as Vade. As Vade explains:

> v002 Rutt/Etra is an attempt to capture some of the beauty of the original hardware. It's almost impossible to emulate an entire Rutt/Etra, as modern graphics and computer systems make fundamentally different assumptions from analog video systems. This plugin is a work in progress, and we will try our best to make it as faithful as possible to the original while exploring new creative possibilities. (Marini 2008)

v002 Rutt/Etra works in the software Quark Composer, a Mac-based visual programming language. Also, reportedly as a tribute to Steve Rutt (the other half of Rutt/Etra), Felix Turner of Airtight Interactive created a still-image 'emulator'. The Rutt-Etra-Izer is WebGL-based software that lets you drop an image into a browser window and change it to resemble an iconic effect (nicknamed the Z-displacement) that was possible with the classic Rutt/Etra (Turner 2011).

Digital preservationists point out what may be obvious: emulators that create digital objects that will also need to be preserved. Of course these emulators are not meant as a means of preservation. What they do is draw attention to the historical tool, and offer a direct way to communicate its concepts and processes. The downside is that the emulators popularize one effect of the synthesizer, and they don't do justice to the very wide range of possibilities of the tool. If emulators were the only form of documentation of a tool, not much would be left of a very complex and influential invention.

A short list of proposals

A studio with an expanded mission: A studio of working electronic tools is the best environment within which to test theories and practices of machine preservation. The Residency program of the Experimental Television Center was aimed at artists, but ETC ran a parallel project in the 'Video History Project', which continues today. Through the 'Video History Project', much information has been gathered that provides a context for the tools and that provides details on their development and operation: oral histories, schematics, operating manuals, manifestos, funding proposals and the like. If ETC finds a home for the studio, the space could both provide artist access and be a center for learning about tool development and tool conservation. Theories of what constitutes a restoration and whether it is possible could be debated and tested. Machines could be analyzed and restored. All of the resources of the various efforts described above could be brought to bear on real, live machines. The home could be an arts center or a museum, but there would be great value in involving like-minded people from science and technology conservation. Just as artists and tool designers of the past drew from the knowledge of each other's disciplines, so could the information that is generated from such a center be useful for conservators of industrial and technology collections, electronic instrument collections and time-based media art.

A survey of video instruments: The exhibition 'Eigenwelt der Apparatewelt: Pioneers of Electronic Art' and the efforts of ETC have done much to locate many of the custom video tools developed in the late 1960s through the 1980s. Research for this book has located even more. An interdisciplinary team, including conservators from the areas discussed above, collectors and technical experts could be created to advise an up-to-date survey. The survey would not just locate the machines, but determine if they are in working condition or 'awaiting restoration'. Informed decisions could then be made about each tool's disposition and further documentation could be collected. The Artists Instrumentation Database, developed for cataloguing machines, could be used to record the information; a simplified version of the template and a cataloguing record for the Raster Manipulation Unit can be found following this essay.[15] Through a survey, the most characteristic tools may surface, and key attributes may emerge that are important to preserve relating to underlying concepts, circuit design or processes. The 'Video History Project' of ETC continues to be a hub for primary documents about custom tool development, and is open to contributions.

Tapping into do-it-yourselfers, inventors and the like: The positive impact of people actively concerned about the longevity of legacy video instruments, especially a community that is networked and engaged, can't be underestimated. Machines, parts and knowledge can be circulated and connections made between generations of tool designers, artists and technical experts.

The idea of DIY has proliferated in technological realms as strongly as any other sector. Art departments are teaching bottom-up device and circuit design. The spirit of the Resurrection Bus – hitting the road in service to media art – and the 1980s Tuesday Afternoon Club – artists gathering to build custom machines – still persists and needs to continue. It is found through a Fixers Collective, in the volunteer staff at a science museum, in online forums, or at video synth gatherings. New ideas for cross-discipline problem solving are needed and can be successful.

Testing the concept of copying: Both industrial and instrument conservation recommend copying for one-of-a-kind or very fragile objects. The original is safeguarded and the new object is used. Copying, where possible, also aligns with the spirit of many of those involved with tool development in the 1970s and 1980s. As Phil Morton said, 'COPY-IT-RIGHT'.

Tapping into a science and technology museum may also be a prospect for copying. Ball expresses enthusiasm for the craft associated with machines that could be useful both for custom electronic tools and for video playback decks, seeing the copying process as:

> [A] marvelous opportunity to recover and demonstrate vanishing crafts [...] the popular view sees a museum's function overwhelmingly in terms of objects. It does not see, or it ignores, the operating, manufacturing and repair skills preserved and sometimes rediscovered by research into working objects. (Ball 1997: 29)

Organizations and schools that are teaching circuit bending and hardware hacking may also be intrigued by the challenge of copying an analog imaging tool.

Document, document, document: Resources like the tools section of the 'Video History Project' and the online archives of the website of The Daniel Langlois Foundation for Art, Science and Technology[16] hold the documents and photographs that fit Ball's idea of enduring evidence of how the tools operated, were used and contributed to art practice. A computer catalog of videotapes, as opposed to a jumble of unidentified tapes, makes for an increased likelihood that an archive will take notice; so does a rich collection of documentation surrounding a custom tool make it more likely that a machine will find an institutional home.

ETC's DVD set *Early Media Instruments* needs to be widely circulated.[17] The laser discs from 'Eigenwelt der Apparatewelt: Pioneers of Electronic Art' contain video documentation of the instruments from that exhibit and associated artists' works. The contents of the discs need to be located, reformatted and uploaded to the 'Video History Project' website. The survey mentioned above and continued documentation go hand in hand. The survey can set a minimum standard for documentation and assess how complete the existing documentation is for a given tool. The information can guide efforts to accumulate as full documentation as possible in physical and web-based archives. A central spot for manuals will be important; the 'Video History Project' website is already functioning in this way.

In the end, theories and proposals only go so far. Custodians will need to continue with their passion and their labor, whether as individuals or as institutions. The progress that has been made on saving 1970s and 1980s noncommercial, independent videotape collections has largely been a matter of the maintenance of ad hoc archives; becoming self-taught conservators and preservationists; making friends in cultural heritage organizations, with technical experts and creative thinkers; and lots of cultural advocacy. Time will tell if the artifacts of this slice of media art history – custom electronic video instruments – will survive through the same care, collaboration and advocacy.

References

Anon (2011), 'Kinect's Grandaddy Running On an Apple IIe In 1978', http://hardware.slashdot.org/story/11/01/27/1945245/kinects-grandaddy-running-on-an-apple-iie-in-1978. Accessed July 30, 2012.

Ball, Stephen (1997), *Larger and Working Objects: A Guide to Their Preservation and Care*, London: Museums and Galleries Commission.

Barclay, Robert (ed.) (1997), *The Care of Historic Musical Instruments*, London: Museums and Galleries Commission; Ottawa: Canadian Conservation Institute, http://www.music.ed.ac.uk/euchmi/cimcim/iht/iht.html. Accessed July 12, 2012.

Cossons, Sir Neal (1999), 'A Perspective on the Nature of Industrial Collections', in Diane Dollery and Jane Henderson (eds), *Industrial Collections: Care and Conservation, Proceedings from the Conference hosted by The Council of Museums in Wales and the United Kingdom Institute for Conservation, Cardiff 9–11th April 1997*, Cardiff: Council on Museums; London: UK Institute for Conservation.

Cycling '74 (2012), 'Video with Jitter', http://cycling74.com/products/max/video-jitter/. Accessed July 7, 2012.

Ditto, Tom (2012), 'Interview with Tom Ditto by Mona Jimenez', July 23, Brooklyn, NY & Ancramdale, NY: Experimental Television Center Archives.

Dunn, David (ed.) (1992), *Eigenwelt der Apparatewelt: Pioneers of Electronic Art*, Linz, Austria: Ars Electronica.

Ecklund, Jon and Richwine, Beth (1993), 'Ensuring a Future for Our Present High-Tech Past: Lessons from the ENIAC for the Conservation of Major Electronic Technology', in *Saving the Twentieth Century: The Conservation of Modern Materials, Symposium Proceedings*, Ottawa: Canadian Conservation Institute, pp. 249–59.

Experimental Television Center (ETC) (2010), 'ETC: 1969–2009', Ralph Hocking and Sherry Miller Hocking (eds), Owego, NY: Experimental Television Center Archives, http://www.experimentaltvcenter.org/etc-dvd.

—— (2011), 'Artists', Owego, NY: Experimental Television Center Archives, http://www.experimentaltvcenter.org/artists. Accessed July 30, 2012.

Gibbon, Richard (1999), 'Controlled Operation or Wrecking? The Use of Objects from the National Railway Museum's Collection', in Diane Dollery and Jane Henderson (eds), *Industrial Collections: Care and Conservation, Proceedings from the Conference hosted by The Council of Museums in Wales and the United Kingdom Institute for Conservation, Cardiff 9–11th April 1997*, Cardiff: Council on Museums; London: UK Institute for Conservation, pp. 17–25.

Hallock, Don and Minkowsky, John (n.d.), 'The Templeton Mixer', http://www.rdlx.com/ncet/mixer/tempmix.html. Accessed July 7, 2012.

Hocking, Ralph (2011), 'Available Machines', Owego, NY: Experimental Television Center Archives, http://www.experimentaltvcenter.org/available-machines. Accessed July 14, 2012.

Jones, David (2005), 'Interview with David Jones by Carolyn Tennant', Owego, NY: Experimental Television Center Archives.

Laurenson, Pip (2007), 'Part 1: Risk Assessment', in *Inside Installations: Preservation and Presentation of Installation Art*, Amsterdam: Instituut Collectie Nederland/Foundation for the Conservation of Contemporary Art, http://www.inside-installations.org/research/detail.php?r_id=83&ct=preservation. Accessed July 23, 2012.

Marini, Anton (2008), 'v002 Rutt/Etra 2.0.1', http://v002.info/plugins/v002-rutt-etra/. Accessed July 24, 2012.

Minns, Jonathan (1989), 'Industrial Archeological Conservation at the British Engineerium', in J. D. Storer and Kenneth Barton (eds), *The Conservation of Industrial Collections*, London: Museums and Galleries Commission and Science Museum, pp. 55–58.

Neal, Blair (2010), 'The Wobbulator', http://blairneal.com/blog/the-wobbulator/. Accessed July 26, 2012.

Odell, Jay Scott and Karp, Cary (1997), 'Ethics and the Use of Instruments', in *The Care of Historic Musical Instruments*, London: Museums and Galleries Commission; Ottawa: Canadian Conservation Institute, pp. 1–7.

Park, Sang Ae and Lee, Ki Jun Lee (2011), *Paik-Abe Video Synthesizer: As Freely As Picasso, As Colorfully As Renoir*, Seoul, Korea: Nam June Paik Art Center.

Rutt, Steve (2007), 'Interview with Steve Rutt by Mona Jimenez', September 6, New York, NY: Experimental Television Center Archives.

Sandin, Daniel (2003), 'Interview with Dan Sandin by Mona Jimenez', May 9, Chicago: Experimental Television Center Archives.

Schlanger, Matt (2012), 'Interview with Matt Schlanger by Mona Jimenez', May 31, Tarrytown, NY: Experimental Television Center Archives.

Summer, Daniel (2007a), 'Interview with Daniel Summer by Mona Jimenez', September 6, New York, NY: Experimental Television Archives.

Summer, Daniel (2007b), 'Interview with Daniel Summer by Mona Jimenez', October 3, New York, NY: Experimental Television Center Archives.

Summer, Daniel (2012), 'E-mail correspondence with Mona Jimenez', July 30.

Turner, Felix (2011), 'Rutt-Etra-Izer', http://www.airtightinteractive.com/2011/06/rutt-etra-izer/. Accessed July 26, 2012.

Waltzman, Mimi S., Barclay, Robert M., Odell, Jay Scott, Karp, Cary and Hellwig, Friedemann (1997), 'Basic Maintenance of Playing Instruments', in *The Care of Historic Musical Instruments*, London: Museums and Galleries Commission; Ottawa: Canadian Conservation Institute, pp. 83–108, http://network.icom.museum/cimcim/resources/the-care-of-historic-musical-instruments-full-text. Accessed November 17, 2013.

WGBH (n.d.), 'New Television Workshop 1967–1999 (1968–1993)', http://main.wgbh.org/wgbh/NTW/FA/COLL/New1.HTML. Accessed June 8, 2012.

Notes

1. For the purposes of this essay, 'instrument', 'tool', 'device' and 'machine' are used interchangeably. The term 'electronic instrument' is used to refer to custom video instruments, not electronic instruments in general, such as drum machines. For a discussion of the use of the term 'instrument' for the custom video tools, see Jean Gagnon's 'Instruments, Apparel and Apparatus: An Essay on Definitions' in this book. References to instruments were not uncommon for designers of custom electronic tools. Dan Sandin resisted labeling of modules on the Sandin Image Processor, saying 'nobody puts the letters on the keys of a piano and nobody puts the letters on a saxophone' (Sandin 2003: 13).
2. For a selection of artists served by the Center, see Experimental Television Center (2011). ETC has produced a DVD set of artworks that were made using the ETC studio: for more information, see Experimental Television Center (2010).
3. Dave Jones's most recent inventions can be seen at Dave Jones Design: http://www.djdesign.com/.
4. See the story, with photos and drawings, in the section 'The Templeton Mixer' at http://ncet.torusgallery.com/mixer/tempmix.html.
5. Technically, the program did not have the name New Television Workshop until 1974. However, the same basic structure – artists using television studios for experimental video – was in place in 1972 (WGBH n.d.).
6. In addition to building a Paik/Abe as a student, Grace later traveled with Paik and Abe, building and trouble-shooting synthesizers. See Kathy High's essay 'Mods, Pods and Designs: Designing Tools and Systems' in this book. Grace is one of a handful of women who were tool builders in the 1970s and 1980s.
7. This summary of the restoration work is drawn from the catalog Park and Lee (2011) and the website for the center: http://www.njpartcenter.kr/en/.
8. The Vasulka online archive can be found at http://www.vasulka.org/.
9. The collection at the Cinémathèque québécoise was formerly held at The Daniel Langlois Foundation for Art, Science and Technology, also in Montreal. At the time of this writing, many collection documents and research guides were still being hosted on the foundation's web site: http://www.fondation-langlois.org/html/e/.
10. Jitter is a product of Cycling '74 that is used for real-time video processing. See http://cycling74.com/products/max/video-jitter/.
11. There are a number of organizations for caretakers of industrial, technology and science collections, including the International Committee for the Conservation of the Industrial Heritage (TICCIH) (http://www.mnactec.cat/ticcih/industrial_heritage.htm) and the Association for Industrial Archaeology (website: http://industrial-archaeology.org/). See also, the Computer Conservation Society (http://www.computerconservationsociety.org/).
12. See Laurenson (2007) for a decision-making process for preserving machines and equipment within time-based media installation.
13. In 2009–11, the Flemish arts organization PACKED and the Netherlands Media Arts Institute collaborated on an Obsolete Equipment Project; see http://nimk.nl/eng/obsolete-equipment. See also Pip Laurenson (2005), 'The Management of Display Equipment in Time-Based Media Installation', London: Tate Gallery of Art.
14. See http://www.lzxindustries.net/.
15. The cataloguing example for the Raster Manipulation Unit is modified from the Artist Instrumentation Database Template. The original template and description can be found on the website of The Daniel Langlois Foundation for Art, Science and Technology http://www.fondation-langlois.org/html/e/page.php?NumPage=710.
16. See the 'Video History Project', http://www.experimentaltvcenter.org/video-history-project; and The Daniel Langlois Foundation for Art, Science and Technology http://www.fondation-langlois.org/html/e/.

A Catalog Record for the Raster Manipulation Unit

Mona Jimenez

Raster Manipulation Unit

Collection of the Experimental Television Center

Date Device Was Invented	c. 1972
Date Machine Was Manfactured/Built	c. 1979

Alternate Name(s)

Wobbulator
Scan Modulator

Inventor/Designer		Role:	
Other Production/Design Credits	Richard Brewster	Role:	Builder; designer for SONY 760 version

Manufacturer or Builder		
Richard Brewster	Place of Manufacture	Binghamton NY
Walter Wright	Place of Manufacture	Binghamton NY
David Jones	Place of Manufacture	Binghamton NY

Extent

1 altered television with 1 mono audio amplifier, 1 stereo audio amplifier, 1 input unit for audio signals, a toggle power switch, a terminal/barrier switch, wires and cables.

Summary Description

The Raster Manipulation Unit is a "prepared television" made from a SONY 760 black and .white television set with a 9" screen. This machine is one of several made at the Experimental Television Center (ETC) starting in the early 1970s based on principles defined by Nam June Paik's early television experiments such as *Dancing Patterns*. This Raster Manipulation Unit was built by Richard Brewster, David Jones and Walter Wright. The devices were nicknamed "Wobbulators" by Walter Wright.

Television sets have deflection yokes that under normal operations guide a stream of voltages in a predictable pattern—in lines from left to right and top to bottom—to create a conventional image. The process of "writing" the lines is referred to as scanning and the resulting pattern of lines on the TV screen is called the raster. The Raster Manipulation Unit, by adding several additional yokes on top of the original ones and allowing the application of various electrical currents (or no current) to one or more of the yokes, changes the conventional scanning pattern and causes image distortions.

The Raster Manipulation Unit takes a video input from a camera or tape, and has presets that allow the user to reverse the image or to collapse the image to a single horizontal line, vertical line or point. In addition, audio signals passed through audio amplifiers can result in the image shifting to the left or right or up or down. There is also a preset that causes a "S" curve in the image. or a rotation of the image around a central point. Any of the effects can also be used in combination, and effects can impact portions of the image or the whole image. In addition the distorted image can be "locked", creating a still image. The Raster Manipulation Unit does not have a video output, so still or moving images are captured through the use of a camera pointed at the TV screen.

The Raster Manipulation Unit was significant in a number of ways. With the Raster Manipulation Unit, one could produce results that could be to some degree controlled and repeatable. In addition, the device marked the beginning of a vocabulary about image processing using external controls, the illusion of a 3D space and the concept of instrumentality. Nam June Paik helped to popularize concepts of raster manipulation; however, the principle of manipulation of the video scanning process was practiced by many artists and inventors and is present in other electronic tools such as the Rutt/Etra Video Synthesizer.

All of the Center's units were made from the SONY 760 set. After the initial Raster Manipulation Unit was built, Brewster designed simple circuitry for that television that helped stabilize the signal. Between 1978 and 1980, Sherry Hocking and Brewster with assistance from Walter Wright wrote documentation for operating and building the device. The documentation was funded by the New York State Council on the Arts and the National Endowment for the Arts. This unit was still in operation in the ETC Studio in 2011.

Device Type: customized/off-the-shelf

Operational Status: operational

Components Description

The major component is a customized television SONY Model 760 9" black and white television set. The set has been disassembled and expanded so that its deflection circuits are exposed. Two deflection yokes have been added on top of the original ones: one yoke from a color television monitor and one wound from enameled copper wire covered with black tape that is referred to in the documentation as the "continuous wind yoke". (Paik used a different design in his raster manipulation units; however, Sherry Miller Hocking remembers him saying that to properly wind a copper wire coil for a prepared TV, one should count to 100 in Korean.)

The television is mounted on a piece of wood. The normal power/volume knob, the UHF and VHF switches and the channel select have been removed or are inaccessible. The brightness and contrast knobs on lower front of the set remain and are functional. The normal connectors on the back of the set are unaltered but are not utilized. Mounted to the wood to the left of the set is a toggle on/off power switch.

There are two BNC connectors mounted on the top of the chassis that are bridged composite video inputs. Behind them is a mounted BNC connector for composite video out (functionality unknown). Next to the video connectors are two mini phone jacks for analog audio in (front) and audio out (back) (functionality unknown). Next to the audio jacks are two black knobs. The right knob controls horizontal deflection of the raster and the left knob controls vertical deflection of the raster. Both knobs have three positions (right to left) normal, reverse, collapse. Reverse flips the raster 180° from normal. Collapse causes all of the lines to be written in one place—a single vertical line in the case of the right knob, and a single horizontal line in the case of the left knob. Used together the raster can be collapsed into a single point.
The audio input unit is a metal mount to the left of the TV. There are three mini phone jacks for input of the control voltages or other external audio signals; these signals are fed into amplifiers. Above the jacks are three black knobs that serve as gain controls for the signals going to the inputs of the amplifiers. They are labeled (left to right) VERT, "S" and HORIZ.

Mounted on the wood to the right of the picture tube is a six position terminal/barrier strip with wires connected to the outputs of two amplifiers. The amplifiers provide strong signals to the deflection yokes. A 60 watt McIntosh 60 mono tube amplifier provides current to the hand wound coil which is adjusted through the "S" gain control. The magnetic fields created by this coil rotate the raster around the center of the image, creating S curves. A Kenwood Model KA -5700 stereo amplifier provides current to the color yoke which is adjusted through the VERT and/or HORIZ controls. These controls allow the raster to be shifted up and down (VERT) or left and right (HORIZ). As noted above, the control voltage inputs can be used alone or in combination, and also with the normal, reverse and collapse functions. Frequencies of control voltages used with this device are generally between 30 Hz and several hundred Hz.

Related Technical Documents & Texts

Hocking, S., R. Brewster and W. Wright. *Raster Manipulation Unit: Operation and Construction.* (online) http://www.experimentaltvcenter.org/raster-manipulation-unit-operation-and-construction Last accessed 5 June 2012.

Sjolander, T. and B. Wilkstom. *Raster Scan Device*. 1966-69. Last accessed June 5, 2012. http://www.experimentaltvcenter.org/ture-sjolander-and-bror-wilkstrom-raster-scan-device

Related Audiovisual Documentation

Neal, Blair, 2010. "The Paik Abe Raster Manipulation Unit a.k.a. The Wobbulator" [video] Availablr at http://vimeo.com/16906546 [Accessed 31 July 2012.

Early Media Instruments: Raster Manipulation Unit 8 DVD set. (Experimental Television Center, 2010))

General Note

In "Raster Manipulation Unit" (see above in Related Documents) the original unit included a monaural amplifier, a stereo amplifier and three audio generators. Also at the time of cataloging, the audio input box was being bypassed.

Copying-It-Right: Archiving the Media Art of Phil Morton

Jon Cates

In 2007 I initiated the Phil Morton Memorial Research Archive in the Department of Film, Video, New Media and Animation at The School of the Art Institute of Chicago (SAIC), through a generous donation from Morton's surviving partner, the late Barb Abramo. This archive contains Phil Morton's 'personal video databank' which stretches across 30 years of media art histories that are specific to the early video art and new media art communities in Chicago. These communities were organized around shared ethical commitments and theory-practice, one of those being Morton's COPY-IT-RIGHT ethic.

During the early 1970s, the work of media artists in Chicago anticipated and developed open-source approaches to free culture, foregrounding collaborative experimentation. Phil Morton's COPY-IT-RIGHT ethic motivated the early video art communities in Chicago and beyond to share resources, widely distribute media and create transparent, decentralized, and open systems.

In 1971, Dan Sandin developed the Sandin Image Processor (IP), a patch-programmable analog computer optimized for video processing and synthesis. Morton, who was a friend and neighbor, asked Sandin if he could build the first copy of Sandin's original Image Processor, and Sandin and Morton began to work together creating the schematic plans for the IP. They named this document the 'Distribution Religion'.[1] Through this philosophy, Sandin open-sourced his unparalleled innovation, giving the plans away for only the cost of the Xerox copies and postage, while simultaneously incorporating any additions or modifications made by those who built their own Image Processors into any further releases of the 'Distibution Religion'. This proto-open-source project gave an international community of artists unprecedented abilities to process and perform real-time audio and video projects.

During this time, Morton developed COPY-IT-RIGHT, an anti-copyright approach to making and freely sharing media art. The 'Distribution Religion' and Morton's individual and collaborative media art works were released under his COPY-IT-RIGHT license. COPY-IT-RIGHT encouraged people to make faithful copies, and to care for and distribute media art works as widely as possible. A close-knit community of collaborators, including Phil Morton, Dan Sandin, Tom DeFanti, Bob Snyder, Jane Veeder, Jamie Fenton and others, worked together in Chicago on the new media of their time, incorporating digital and analog computing with real-time audio and video synthesis, processing, programming and performance. Many of these individuals, as well as Morton's students, adopted the COPY-IT-RIGHT ethic, applying and modifying it in their own projects. (See Color Plate 27.)

Schools such as SAIC and The University of Illinois at Chicago – Circle Campus acted as incubators for the ideas and approaches that these early media art communities shared. Students and faculty formed new departments and organizations that would develop into internationally recognized homes of artistic experimentation and technological development. At SAIC, Phil Morton founded the Video Area and the Video Data Bank. During this same time period, Dan Sandin and Tom Defanti founded the Electronic Visualization Lab at the The University of Illinois at Chicago – Circle Campus. The faculty and students of these two schools and their freshly founded departments worked together informally and publicly to build various forms of inter-institutional collaboration.

Phil Morton arrived in Chicago in 1969, joining the Faculty at SAIC. Only one year later Morton founded the Video Area, which was the first department in the United States to offer a BA and MFA degree in video art. Within the Video Area, Morton founded The Video Data Bank, which has now become one of the world's leading collections of video art. The Video Area later became the Video Department, which then merged with the Film Department to form the Department of Film, Video, New Media and Animation. Frequent guests in the Video Area during the 1970s included Steina and Woody Vasulka, who established The Kitchen in New York City, and Gene Youngblood, who published *Expanded Cinema* in 1970. Graduate students such as Lyn Blumenthal, Kate Horsfield, Christine Tamblyn and Jane Veeder helped to form the early video art community at SAIC, and also became important contributers and developers of the early culture of media art in Chicago.

In 1969 Dan Sandin was, in his own words, 'hired to bring computers into the art curriculum' (criticalartware 2003a) of the University of Illinois at Chicago Circle Campus, then referred to as Chicago Circle Campus. Tom DeFanti (originally hired as faculty in the Chemistry Department) joined Sandin and they co-founded the Circle Graphics Habitat at Chicago Circle Campus in 1973. The Circle Graphics Habitat would soon be renamed the Electronic Visualization Laboratory (Electronic Visualization Lab), and would continue functioning into the present at the leading edge of the intersections of art and engineering. When DeFanti joined the faculty at Chicago Circle Campus and cofounded the Electronic Visualization Lab with Sandin, he brought with him the Graphic Symbiosis System (GRASS) that he had developed as a PhD student at Ohio State University while working with Charles Csuri, a pioneer in the field of computer art (Rivlin 1986: 15). GRASS was a real-time computer graphics programming system that allowed for real-time interactions, allowing programs to be written or altered interactively while the visual graphics output of those programs was rendered on-screen. During this time, Ted Nelson, who is famous for his coining of the terms 'hypertext' and 'hypermedia', also joined the faculty at Chicago Circle Campus. Nelson was roommates with DeFanti and self-published his now famous *Computer Lib/Dream Machines* (1987) in 1974 while living in Chicago with DeFanti.

During the highly active period in the 1970s, artists, educators, innovators and organizers paths crossed and became deeply interwoven in Chicago, forming the closely knit fabric of these media art communities. These individuals included those already mentioned in this text and others to form an impressive set of pioneers: Dan Sandin (who additionally cocreated the CAVE Virtual Reality system); Tom DeFanti (cocreator of CAVE Virtual Reality system); Bob Snyder (founder of the sound department at the SAIC); Jane Veeder (early digital artist recognized by 'Ars Electronica' and SIGGRAPH); Jamie Fenton (cofounder of MacroMind, which would become MacroMedia; computer programmer and developer of the software that would become Adobe Director); Larry Cuba (accomplished media artist who contributed to early computer animation); Ted Nelson (who in addition to hypermedia and hypertext also invented the concept of Xanadu); Timothy Leary (the countercultural futurist and provocateur); Gene Youngblood; Steina and Woody Vasulka (media artists who importantly contributed to early video art); Sonia Landy Sheridan (whose work on generative systems at SAIC has been documented by Kathryn Farley for The Daniel Langlois Foundation); and Kate Horsfield and Lyn Blumenthal (who took over the direction of the Video Data Bank at SAIC and transformed the organization into a world leader of video art collection and distribution). The stories and histories of those individuals and their achievements have all been fairly well documented and incorporated into media art histories; however, Phil Morton's work – which literally connected and brought into conversation many of those above – has remained little discussed.

Phil Morton's absence triggered my interest in this alternative media art histories project. I was familiar with the work of many of the above-mentioned artist-educators but mostly unfamiliar with Morton and gained access to some of his work through my position in the Department of Film, Video, New Media and Animation at SAIC. As I watched and read through the materials that I unearthed, I became excited and surprised by Morton's work and wanted to see as much as possible. Uncovering this material, I began to wonder how this aspect of media art history had been lost, forgotten or repressed in the media art historical account in general and also within SAIC's institutional memory.

I have been able to theorize a few of the reasons for this situation. Morton vehemently advocated for free culture and open-source approaches to media art before such ideas were understood or such terms were in use. He experimented relentlessly with boundaries, ignoring as many distinctions between personal, professional, political, aesthetic and technological categories as possible. He quickly moved to include analog and digital computing in his artistic work and academic curriculum with very few antecedents to rely on or refer to. In doing so, he purposefully and playfully explored what we would now refer to as new media art, an art that was radically open, remixed, collaborative and conversational. As such, his projects were innovative and anticipatory as well as hybrid in form, difficult to categorize, anticommercial, anticorporate and sprawling in their cyber-psychedelic sensibilities.

Both having arrived in Chicago in 1969, Morton and Sandin met on Chicago's South Side, in the Pilsen neighborhood where they had neighboring artist studios. Morton and Sandin became friends and collaborators in 1971 while Sandin was in the early stages of developing his IP. As Sandin explains, the IP is a 'general purpose patch-programmable analog computer [...] optimized for processing video information' which can also 'process sound or any other signal' (Sandin 1973). By 1973, when complete, fully documented and copyable, the Sandin Image Processor offered artists new abilities to create, control, affect and transform video and audio signals in real time, enabling live audio-video performances and the production of work that would become collectively known as 'image processing' and associated with the early media art community in Chicago. 'Image processing' is a subgenre of video art historicized and documented in detail in the 'Performance of Video-Imaging Tools' section of *Surveying the First Decade: Video Art and Alternative Media in the U.S.* (1995), the Video Data Bank's historical collection of video art. As Chris Hill writes in an essay that accompanies the video collection, an interest in developing systems and tools, such as the IP, was shared among many of those in the video art movement across the United States. Hill writes that 'Throughout this period, artists, usually in conjunction with independent engineers, modified and invented video "instruments" or imaging tools, making possible the construction of new video and audio systems shaped by their individual aesthetic agendas' (Hill 1996). The IP is one of the most well known and well documented of these instruments.

In 2003, Mona Jimenez developed 'The Artist Instrumentation Database Project' as a researcher-in-residence with The Daniel Langlois Foundation (Jimenez 2005). The Artist Instrumentation Database Project is among recent studies of the activity of instrument building in the early video art communities of the United States. Jimenez compared two custom-built tools developed and used by artists during this period, referring to 'technological devices as art-making tools and as components of electronic art works' thus linking the artist-inventors' work of technological innovation in the creation of these instruments with their placement within works of art (2005: 1). If these devices are developed by artist-inventors and used not only to make art but as components of electronic art works, then they may also be considered, at least in part, as art works in and of themselves. The two instruments that Jimenez details in her research are the Rutt/Etra Scan Processor and the Sandin Image Processor. As Jimenez states, these two instruments were actively used in the 1970s and 1980s in and for the production of 'time-based media art and performance incorporating media' (2005: 1). Jimenez recounts that both instruments were inspired by previous developments in media art and electronic music, such as Nam June Paik and Shuya Abe's Paik/Abe Video Synthesizer and Bob Moog's audio synthesizers. Still, each artist-developer of video instruments took fundamentally different approaches to the conceptual and technical structures of the instruments and the cultural or commercial distribution of these devices or systems.

The IP is an open system of which 'numerous versions were built' (Jimenez 2005: 3). As Jimenez has documented, those that copied the Sandin Image Processor undertook this work themselves, often extending and expanding upon the functionality of the system. This kind of versioning, extension and expansion was possible due to the 'early model of open source' (Jimenez 2005: 1) that Sandin used to distribute the system. Sandin himself also notes a similarity to, and anticipation of, open-source software in his approach to the distribution of the Image Processor, saying, 'it is very much like the open-source development that we have today in software' (criticalartware 2003a).

The IP became open source because of Morton's interest in making the first copy of the IP. Consequently, the documentation of the system that Sandin and Morton developed together (in order for Morton's first copy to be created) made the wider media arts community aware of the IP. This documentation is known as the 'Distribution Religion'. Morton proudly referred to himself as the 'first copier' of an IP (Morton 1973: 2) and his first act of copying enabled an entire community to develop. As Sandin recalls in a 2002 interview with criticalartware, when Morton first asked Sandin if he could copy the IP, Sandin said yes but was not sure it was technically possible:

> [Y]ou definitely have permission to build a copy, but I actually don't know if you can build a copy. I didn't want to be insulting, but I didn't really know what it took, because you're so immersed in this stuff, in designing and building it, it's hard to figure out what resources are actually necessary to do a copy. We talked about it a little bit and he built the first copy. As part of that process of building the first copy we did a documentation that was sufficient so that other people could copy it. It took a year's worth of Friday afternoons where I'd show up at Phil's house and we'd work on his IP for a while and he'd produce part of the documentation and I'd work on it. We developed a format and a way of doing it. I won't say it was intensive work every Friday afternoon but it took a year of getting together and fiddling at it. By the end of that we actually accomplished a document which enabled a large number of people. In a period of a few years, twenty or so copies of the IP were built, almost all of them by individuals or small arts institutions. (criticalartware 2003a)

'Distribution Religion' contains all of the schematic plans and electronic diagrams necessary to build an Image Processor in its 'classic' configuration, and also documents any significant contributions made by those who extended or modified the system. Thus, the 'Distribution Religion' not only anticipated artists' current approaches to free and open-source software development, but also distribution methods and do-it-together (DIT) collaborations.

Although primarily a technical document, 'Distribution Religion' also offers articulations of the techno-social and media-philosophical positions of Sandin and Morton. These philosophies introduce and conceptually enframe the technical plans. In the introduction, Sandin wrote: 'I think culture has to learn to use high-tek machines for personal aesthetic, religious, intuitive, comprehensive, exploratory growth. The development of machines like the Image Processor is part of this evolution' (Morton 1973: 1). Morton goes on to write 'NOTES ON THE AESTHETICS OF "copying-an-Image Processor"', saying that he is a 'copier of many things' and 'the first copier of an Image Processor' so that his notes and observations that follow should be meaningful to 'future copiers' (Morton 1973: 2).

Morton wrote with emphatic enthusiasm that:

> First, it's okay to copy! Believe in the process of copying as much as you can; with all your heart is a good place to start – get into it as straight and honestly as possible. Copying is as good (I think better from this vector-view) as any other way of getting 'there'. (Morton 1973)

Morton articulated an ethical position on the act of copying in 'Distribution Religion'. He holds that copying is right, morally correct and good, and in the case of the IP, necessary for its distribution as an instrument that was conceived of (conceptually and technically), through an implementation of Morton's concept of COPY-IT-RIGHT, as expandable, decentralized, free and open source.

The IP is a system that is expandable or extendable via the creation of new modules, components and plug-ins developed by the community of codevelopers who reimagined the possibilities of the system. Thus structurally Sandin's system is open and decentralized; the decentralization of the system echoes a technical, structural and philosophical position outlined in 'Distribution Religion'. Christine Tamblyn states in her history of image processing (as a subgenre of video art made in Chicago) that 'Sandin eschewed the hierarchical organizational mode favored by television engineers that places all components under the control of one centralized unit' (Tamblyn 1991: 304). The classic configuration of the IP allows all basic functions to operate, but the modules can be arranged and rearranged to achieve highly customized effects and personalized styles, to attain what Tamblyn calls 'unpredictable and unprecedented results' (1991: 304).

Many of Morton's projects relied heavily on the Sandin Image Processor, as can be seen online via the 'Phil Morton Memorial Research Archive', where Morton's works are freely available for viewing and download.[2] The Archive contains over five hundred videotapes from the early 1970s and 1980s in U-Matic tape, ½" open-reel and VHS formats, as well as videotapes from the 1990s and associated paper files, notes, and other documentation. Many of the videotapes are process-oriented, consisting of raw materials, various versions, final edits and subsequent remixes. The availability of the process materials is what makes the Archive particularly useful for research. Morton's process from initial acquisition of footage through various forms of image processing of the footage and its incorporation into edited versions, final edits and remixes can be studied.

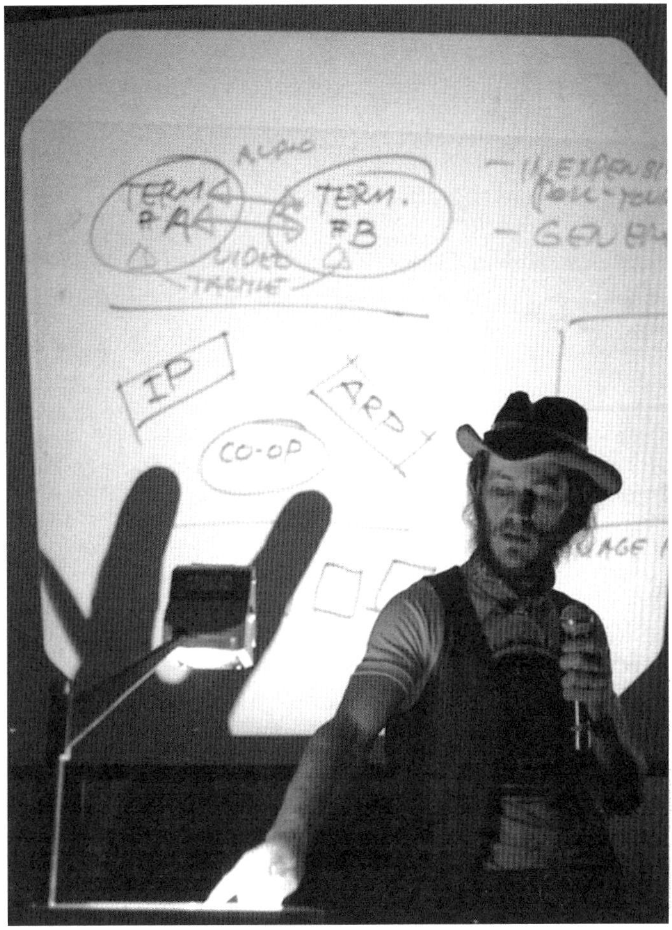

Figure 1. Phil Morton lecturing at 'Design/Electronic Arts', a conference in Buffalo, New York (1977). (photo & courtesy. Jane Veeder).

Much of Morton's material is collaborative, involving projects with Sandin, DeFanti, Snyder and Jane Veeder. The collaborative work is almost always explicitly positioned as anti-copyright, as it is released under Morton's COPY-IT-RIGHT ethic in much the same manner as the release of 'Distribution Religion'. I will now contextualize Morton's principle and practice of COPY-IT-RIGHT in relation to current forms of resistance to copyright and intellectual property regimes. The continuum of alternatives to copyright provides a set of contemporary comparisons to COPY-IT-RIGHT and serves to illustrate the ways in which Morton's COPY-IT-RIGHT ethic anticipated, and has several different relationships to, these current models.

As noted above, Morton used his anti-copyright COPY-IT-RIGHT ethic, or what one could call his licensing system, for his individual and collaborative media art works during the period when he was developing his IP and the 'Distribution Religion'. 'NOTES ON THE AESTHETICS OF "copying-an-Image Processor"' appears to be the first fully formed articulation of Morton's COPY-IT-RIGHT ethic. In the years that followed, Morton would continue to define COPY-IT-RIGHT through his writings and his collaborative media art projects. By the late 1970s and early 1980s, his COPY-IT-RIGHT ethic would be visibly and fully articulated in his works. Furthermore, his students, colleagues and collaborators would use this ethic during this time period, taking up his collaborative and anti-copyright approaches to making and distributing experimental media art. Not only did they share their projects with one another, but in the folk technology tradition initiated and promoted by 'Distribution Religion', they openly shared their methods for the production of their work and came to feel strongly that their work was operating simultaneously as media art and community building.

Morton's former artistic collaborator and partner Jane Veeder explains Morton's COPY-IT-RIGHT ethic:

> Phil was really famous at the time for copying everybody's work, he was always famous for copying people's tapes. There was that early counterculture sense, and it's strong in the digital realm as well, that counterculture sense that information should be free. Now it's actually gaining momentum. Look at the open-source software development movement. Look at Linux. A lot of people find fault in Linux but nevertheless, it's certainly got a lot farther than people thought at the time, and I am sure there are many that are threatened by it. Phil had an expression that was, 'copy it right'. The idea was to make a faithful copy, take care of it, show it to people, and that justified making a copy of anything. (criticalartware 2003b)

Veeder was Morton's closest collaborator and partner during Morton's most artistically active period – the mid to end of the 1970s. Not only did they collaborate on experimental media art projects, they started new academic initiatives and media art organizations at SAIC, and traveled the United States together. Veeder's comments purposefully make it clear that Morton's stance could be thought of not only as anti-copyright and open source, but also as pro-piracy. The ethic of COPY-IT-RIGHT thereby emphasizes a multivalenced moral imperative to freely copy and openly distribute creative works as widely as possible, especially in the hostile environment of copyright and intellectual property regimes that would make such activities illegal.

My use of the phrase 'intellectual property regimes' points towards the techno-social roles of archives. Should archives function primarily as collections of protected intellectual property assets, locked and gated away behind layers of copyright protection; or should archives function as available resources that can allow materials to be further extended and engaged with in open and unforeseen ways? While these are simplified options, with the Phil Morton Memorial Research Archive my task is to facilitate the latter of the two. That is to say, the archive must not become another indexed and cataloged set of overly protected intellectual properties in a library science–style approach that would seal and stabilize the ownership of these media objects, but rather the objects must remain as they were intended to be when created – used, reused and understood as artistic dialog in processes of constant becoming.

As Janice T. Pilch states in her essay 'Collision or Coexistence? Copyright Law in the Digital Environment', intellectual property regimes are most 'often associated with efforts to wipe out music and film piracy' (2005: 80) such as those made possible by file sharing and copying of digital files online or in peer-to-peer networks. Morton's COPY-IT-RIGHT ethic and his formation of the Video Data Bank – which was initially designed as an open collection of copied and copyable resources – were conceived of to support digital sharing and distribution of media arts, thirty years before the possibility of exchanging video materials in, on or through online networks and personal computing devices. Even the globalized intellectual property regimes that exist today have been defined in the United States and filtered through a great deal of transnational corporate interest.

The current copyright laws began to be revised and firmly established during the 1980s through The General Agreement on Tariffs and Trade (The GATT); in the 1990s through the formation of the World Trade Organization (WTO) and the WTO's Agreement on Trade-Related Aspects of Intellectual Property Rights (the TRIPS Agreement); and most recently the Digital Millennium Copyright Act (DMCA). The development of these laws, legal systems and organizations are generally understood to have been landmark events in the process of globalization. The DMCA is the most digitally specific, wide-reaching and expansive of these laws, and has recieved widespread criticism for its severity, scope and enforcement.

Various forms of resistance to copyright have been identified by scholars such as Debora Jean Halbert in her work on intellectual property. Halbert seeks to find and highlight the strengths of 'alternatives to protecting knowledge resources that don't translate them into private property' (2005: 5) while investigating a number of areas in the legal expansion of copyright; her focus is on the ways in which intellectual property regimes limit creativity and the exchange of information while increasing suspicion among creators and distributors of content. One of the failures of imagination in the current globalized legal system of copyright and intellectual property is, in Halbert's words, the assumption that 'creation stems from the chance of monetary rewards' (2005: 5). Morton and his Chicago-based group of collaborators and students did not share this assumption, and considered their creative work to be for the moral, artistic, personal and political good of their communities, where they represented a formative core of artist-educators. As such, Halbert's search for alternatives and resistance to, as well as critiques of, copyright law and intellectual property regimes hold particular importance, underlining that the experimental work undertaken by these artist-developers is echoed in critical and scholarly analysis thirty years later. Over the course of these thirty years, the issues of copyright and intellectual property in media arts become even more pressing, as the digital forms that Morton and his collaborators developed and experimented with eventually became the basis by which almost all media is rendered, distributed and exchanged.

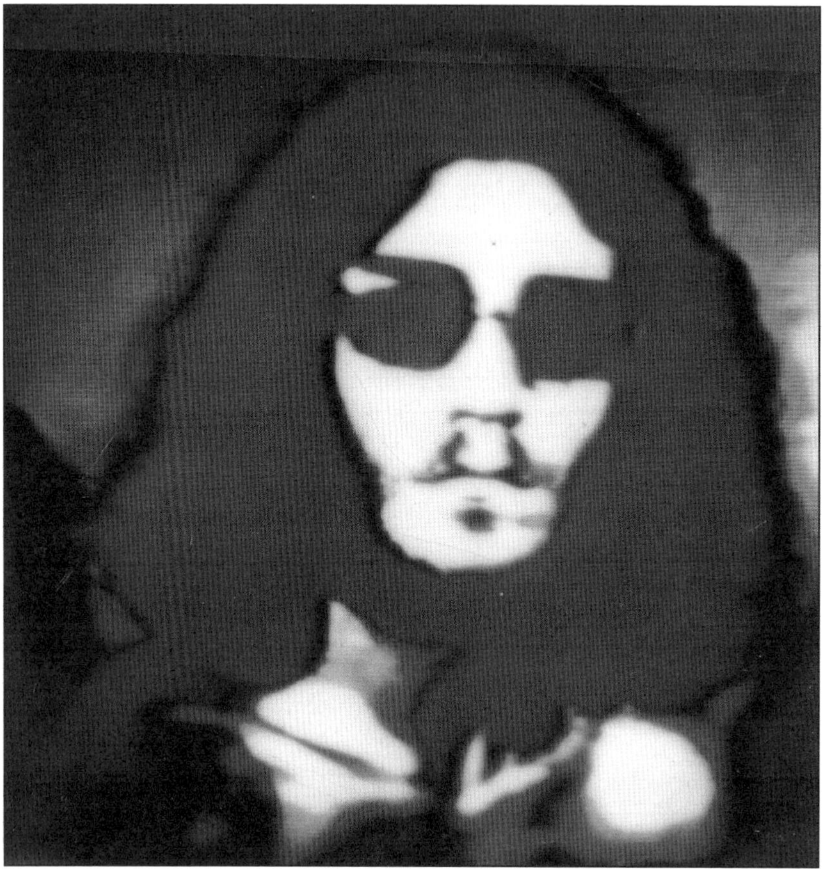

Figure 2. Phil Morton self-portrait. (courtesy. Phil Morton Memorial Research Archive).

Halbert notes that since the 1990s, with the commercial popularization of the Internet through the development of the World Wide Web and the availability of digital video, the United States Congress and the intellectual property industries (including the transnational corporations that own the rights to creative works, the Motion Picture Association of America, the Recording Industry Association of America, their legal teams, etc.) have responded by legally extending the concept of copyright 'as it confronted the new technologies of the information age' (2005: 1). These changes to the idea of copyright and the laws governing copyright and intellectual property articulate a 'desire to own what only has value through circulation and to control every possible exchange of this information' (2005: 3). Halbert refers to the resulting increase in suspicion 'as everyone starts worrying about property and not about sharing the results of their intellectual or creative work'

(2005: 3). Key to her critique of copyright and intellectual property law is that the increased suspicion is corrosive to creative communities.

Halbert suggests that the worry or suspicion is not irrational or unwarranted, and that alternatives need to be developed and explored. Halbert explains that as 'solutions become increasingly draconian with each new lobbying round by major intellectual property interests' (2005: 3), and the conceptual framework of property is the main way in which creative work is enframed or understood, more suspicion is produced. This suspicion has a destructive effect, causing people to worry about 'how their work will be misused instead of used' (2005: 3). Halbert continues, writing that '[c]oncerns about property protection do nothing to enhance the free exchange of ideas' (2005: 3). Rather than promoting a culture in which the creative arts are valued in frameworks other than property, and artists are encouraged to freely exchange and share ideas, copyright law as enacted by the United States Congress has, according to Halbert, moved to 'provide even stronger protective measures for copyright and patent owners' (2005: 3). More directly, the United States Congress has enacted laws at 'the behest of the entertainment industries' (2005: 3) in order to further expand the definitions of copyright and intellectual property in favor of the industries, and in order to further protect these corporate interests rather than the public interest. In particular, the United States Congress and the corporations that this legislative government body has defended and acted at the request of, drafted the DMCA in part to respond to new digital technologies, but also as an opportunity to expand the definitions of copyright and intellectual property for the sake of corporate copyright owners. Halbert plainly states that these combined facts demonstrate that the DMCA 'clearly illustrates that the law is not a neutral body of abstract principles, but is instead the codified will of those with economic and political power' (2005: 3).

Matteo Pasquinelli has written on these issues in *The Ideology of Free Culture and the Grammar of Sabotage*. Pasquinelli analyzes alternatives to current copyright law such as the Creative Commons initiative. He states that Creative Commons faces 'a growing criticism that comes especially from the European media culture' (2008). The European media culture critics cited by Pasquinelli include Florian Cramer, Anna Nimus, Martin Hardie and Geert Lovink. Among the critiques, Pasquinelli defines two main positions. The first focuses on the fact that the producer-centered ethic of Creative Commons does not recognize or include a critical rethinking of the uses of media produced under a Creative Commons license and therefore continues to contribute to a sociopolitical imbalance in the techno-social creation/construction of 'producers' and 'consumers'. The second position highlights that Creative Commons is consistent with existing copyright laws and therefore does not provide a real alternative. Pasquinelli advocates instead that 'a tactical notion of autonomous commons can be imagined to include new projects and tendencies against the hyper-celebrated Creative Commons' (2008).

To imagine an autonomous commons, Pasquinelli refers to the concept of 'Copyfarleft' by Dmytri Kleiner. 'Copyfarleft' opposes systems of private control over the means of publication, distribution, promotion and media production. This opposition takes the form of a critique of ownership of material assets, recognizing class issues and allowing workers to reclaim production. Copyfarleft ensures that products such as media art works remain free, and as Pasquinelli states, 'can be used to make money only by those who do not exploit wage labour (like other workers or co-ops)' (2008: 7). This explicitly ethical sociopolitical position connects back to the concept of COPY-IT-RIGHT, as Morton's position was also an 'ethic' that opposed private property, ownership and economic exploitation on the basis of new technologies.

Also, Florian Cramer has addressed the ways in which current copyright and intellectual property laws are not neutral, and has suggested that alternatives need to be explicitly articulated in ethical terms. Furthermore, Cramer has consistently addressed these concerns to new media art communities through his writings and presentations at festivals such as 'Ars Electronica' and 'Wizards of OS', and his posts to lists such as nettime. Cramer advocates for new media artists to critically analyze the context of current copyright laws as they relate to creative and computational works. Cramer has detailed how the General Public License (GPL) and free and open-source software movement's strategies can be applied to more than simply software, and how these and other alternatives to copyright relate to media art.

As Cramer explains, the General Public License differs significantly from that of Creative Commons. Creative Commons has gained popularity among new media artists and in media art communities, while the GPL has not been as widely adopted as a way to distribute works. While the principle of openness may have popularity among new media artists, Cramer encourages these artists to weigh the differences, and carefully consider the implications and results of using variously open or closed approaches such as those presented by Creative Commons, the Free Software Foundation or the Open Source Initiative. Inke Arns (who has collaborated with Cramer on curatorial projects that address these topics of openness) has written on the art–historical practices and movements that have informed current forms of resistance to copyright. In her 2008 exhibition catalog text 'Use = Sue: On the Freedom of Art in the Age of "Intellectual Property"', Arns traces these histories through situationist detournement in the 1950s, the cut-up method of Brion Gysin and William S. Burroughs in the 1960s, the Neoist 'Festivals of Plagiarism' in the late 1980s, and what she calls 'a broad culture of appropriation' in art movements of the twentieth century such as Pop Art (Arns 2008). In these analog forms, including the film art subgenre of found footage, Arns states that artists engaged in these practices 'undermine concepts like originality and authenticity' and 'subvert the nineteenth-century Romantic concept of the artist-genius autonomously creating from within' (Arns 2008). Placed in this context, Morton's collaborative works and the COPY-IT-RIGHT ethic can be understood alongside these media art–historical moments, offering an available alternative approach.

When discussing the shifts that have occurred in the transition to the twenty-first century, Arns quotes philosopher Eberhard Ortland's analysis of the expansion of copyright law, stating that this situation is 'leading to an unparalleled concentration of resources in the hands of globally active quasi-monopolists in the media and IT markets' (Arns 2008). Arns applies Ortland's critique to the arts. Still, Arns does not directly address media art or include the video art histories in her text. I raise Arns's art–historical account at this point to again underscore that what Arns refers to as the 'Asymmetrical Expansion of Copyright to the Advantage of Exploiters' (2008) is an important, if often overlooked, aspect of media art histories relating to current new media art–theory practices, and free and open-source software cultures. Arns's example also serves to point out that even among those who are working and writing critically in the field of media art (and as someone with whom Cramer has collaborated), the realities of copyright, intellectual property and free and open-source software cultures provide important and often underrepresented critiques and alternatives that can be used in the education of new media artists.

Cramer similarly references art–historical trajectories and motivations for present-day interest in, and commitment to, resisting copyright, but is particularly concerned with the cultural implications of code and new media art as forms of anti-copyright activism. When Cramer compares options for openness in media art, he questions the options available under Creative Commons on the basis of their 'lack of an underlying ethical code, political constitution or philosophical manifesto such as the Free Software Foundation's Free Software Definition or Debian's Social Contract and the Open Source Initiative's Open Source Definition' (Cramer 2006). As he states, the lack of a fundamentally ethical, political or philosophical basis to their options undermines the effectiveness of Creative Commons from the onset. As has been noted above, Morton's COPY-IT-RIGHT concept is primarily an ethical, political and philosophical position, and as such remains on a firm basis if placed into the continuum of Cramer's analysis of comparative forms of openness in media art.

Cramer has defended the application of the Free Software Foundation's General Public License to non-software work; in other words, to artistic practice that is not explicitly or literally software art. The Free Software Foundation released Version 1 of the General Public License in 1989 in order to articulate and defend their position that software should be freely available (to distribute and change, improve or extend) to both developers and end users. The Free Software Foundation clearly maintains that this ability is a 'freedom' which should be 'guaranteed' and maintained in resistance to commercial software companies who 'try to keep users at the mercy of those companies' (Free Software Foundation 1989). As Cramer notes, the only reason that the General Public License has not been more quickly or widely applied to artistic practice is that this license 'speaks of the licensed work as "the program", not "the work"' (Cramer 2005). Still, as Cramer reminds the new media artists, theorists and critics to whom he addresses, 'many non-software works, such as manual pages, have been released under the GPL and continue to be released' (2005).

In the same text, Cramer continues by reiterating that the Creative Commons, despite its popularity, does not actually encourage the development of a commons or shared and freely accessible cultural space. The Creative Commons project is actually intended to maintain coherence with existing intellectual property and international copyright laws. These laws are based on the assumptions of a profit motive for creative work, and an assertion that creative works are private property – identifiable, distinct products made by those who claim authoritative authorship. This situation, which confuses the concept of the Commons with adherence to current intellectual property and international copyright law, according to Cramer, results bluntly 'in a big mess and confusion' (2005). As Cramer concludes: 'Perhaps one should start an advocacy effort for the GPL as the good, intellectually beautiful, standard license for free work' (2005). While Morton's COPY-IT-RIGHT may or may not fulfil the same criteria if it were to be considered by Cramer, COPY-IT-RIGHT certainly does provide a media art–historical example of an alternative to both copyright and similarly influential counter-cultural proposals introduced in the early 1970's such as Ted Nelson's Xanadu with its Creative Commons-like method of adhering to copyright while advocating for expanded flexibility (and commoditization of increasingly smaller transactions) in digital networks.

As has been explained above, Creative Commons is not an alternative or a form of resistance to copyright, but was intended from its inception to be offered as being consistent within existing copyright law and intellectual property regimes. However, the question of the historical perspective of Creative Commons remains relevant to understanding an alternative historical account, and in order to explain the absence (within the framework of Creative Commons) of an understanding of the types of resistance to copyright as have been outlined above. In 2002, Lawrence Lessig defined many of his ideas of Creative Commons in his *The Future of Ideas: The Fate of the Commons in a Connected World*. In 'Creativity in the Dark Ages', Lessig refers to the dark ages before the Internet, which he defines as the 1970s (Lessig 2002: 104). He claims to detail two forms of creativity – in 'The Arts' and in 'Commerce'. His claim to be discussing 'The Arts' is only through a very brief outline of the conditions of mainstream media, in particular the development of cable television in the United States. He does not engage with media art or video art histories that include alternative artist-made media of the late 1960s or 1970s.

This oversight puts Lessig at a disadvantage by limiting his ability to encounter work such as Morton's COPY-IT-RIGHT, which directly critiqued copyright prior to the popularity of free and open-source approaches or of Creative Commons. Lessig imagines Internet technologies to present radically new and unprecedented abilities for exchange, sharing and collaborative or collective authoring of media. The radically new and unprecedented status that Lessig asserts requires a denial of the alternative media art histories considered here. The physical and technologically mediated networks that Morton and his collaborators used to exchange (or in their own words 'bicycle')[3] tapes, and to copy them in order to further redistribute and remix them, is overlooked by Lessig; however, these acts provide an important foundation for current Internet-based, digital and new media art-related systems that are similar in their alternative forms of production, distribution, use, reuse and exhibition.

In contrast, Halbert advocates for an analysis, such as the one this study undertakes, into alternative media art histories, that can provide historical accounts of forms of resistance to copyright in media art cultures and communities. As Halbert writes, transnational corporations with desires for 'monopolistic control over content' have increasingly sought and received legal support for expanded control over creative content and 'the vehicles through which that content is provided', in order to envision a future wherein all content is exchanged within 'the framework of profit', and any other vision of possible futures is 'utopian' (Halbert 2005: 7). As Halbert states (and as the example of Morton's COPY-IT-RIGHT makes clear): 'There are thousands of people around the world who have developed their own ways of dealing with what we call intellectual property' (2005). Halbert concludes hopefully that these 'parallel systems, alternative paradigms, and small resistances prove that we do have a choice in how the future develops' (2005). This future, in which these alternatives to copyright can exist and flourish, relies on a recognition and critical inclusion of underrepresented, repressed, lost or forgotten histories (such as the subject of this study) in order to establish the past upon which a viable future of ideas can be based.

References

Arns, Inke (2008), *Use = Sue: On the Freedom of Art in the Age of 'Intellectual Property'*, Hartware MedienKunstVerein, http://www.hmkv.de/dyn/_data/2008_Arns_Anna_engl.pdf. Accessed July 15, 2012.

Cramer, Florian (2005), 'Re: <nettime> What's the meaning of "non-commercial"?', nettime, http://amsterdam.nettime.org/Lists-Archives/nettime-l-0501/msg00012.html. Accessed July 15, 2012.

—— (2006), 'The Creative Common Misunderstanding', nettime, http://www.nettime.org/Lists-Archives/nettime-l-0610/msg00025.html. Accessed July 15, 2012.

criticalartware (2002), 'Kate Horsfield interviewed by criticalartware', http://www.copyitright.org/criticalartware/KateHorsfield/. Accessed July 15, 2012.

—— (2003a), 'Dan Sandin interviewed by criticalartware', http://www.copyitright.org/criticalartware/DanSandin/. Accessed July 15, 2012.

—— (2003b), 'Jane Veeder interviewed by criticalartware', http://www.copyitright.org/criticalartware/JaneVeeder/. Accessed July 16, 2012.

Free Software Foundation (1989), 'General Public License', The Free Software Foundation, http://www.gnu.org/licenses/gpl-1.0.html. Accessed July 16, 2012.

Halbert, Debora Jean (2005), *Resisting Intellectual Property*, London: Routledge.

Hill, Chris (1996), 'Attention! Audience! Production! Performing Video in Its First Decade, 1968–1980', Video Data Bank, http://www.vdb.org/node/13086. Accessed July 16, 2012.

Jimenez, Mona (2005), 'The Artist Instrumentation Database Project', The Daniel Langlois Foundation, http://www.fondation-langlois.org/html/e/page.php?NumPage=708. Accessed July 16, 2012.

Lessig, Lawrence (2002), *The Future of Ideas: The Fate of the Commons in a Connected World*, New York: Vintage Books.

Morton, Phil (1973), 'NOTES ON THE AESTHETICS OF "copying-an-Image Processor"', criticalartware, http://www.copyitright.org/DistributionReligion.pdf. Accessed March 30, 2013.

Nelson, Ted (1987), *Computer Lib/Dream Machines*, 2nd edition, Redmond, WA: Tempus Books/Microsoft Press. First published 1974.

Pasquinelli, Matteo (2008), *The Ideology of Free Culture and the Grammar of Sabotage*, The Institute of Network Cultures, Rotterdam: NAi Publishers, http://www.rekombinant.org/docs/Ideology-of-Free-Culture.pdf.

Pilch, Janice T. (2005), 'Collision or Coexistence? Copyright Law in the Digital Environment', in M. Neubert (ed.), *Virtual Slavica: digital libraries, digital archives*, New York: The Haworth Information Press, pp. 79–116.
Rivlin, Robert (1986), *The Algorithmic Image: Graphic Visions of the Computer Age*, Redmond, WA: Microsoft Press, p.13.
Sandin, Dan (1973), *5 Minute Romp Through the Image Processor*, Chicago: Video Data Bank.
Tamblyn, Christine (1991), 'Image Processing in Chicago Video Art, 1970–1980', *Leonardo*, 24: 3, p. 304.

Notes

1. The 'Distribution Religion' can be found in this book in Section 3: People.
2. The archive can be found at http://copyitright.org.
3. 'The only way you were able to see anybody's work wasn't through festivals. There weren't any festivals then, it was through what everybody called "bicycling". Somebody would send Phil Morton a tape and he would invite everybody over to his house to look at it' (criticalartware 2002).

Proposal for Low-cost Retrieval of Early Videotapes Produced on Obsolete Equipment and/or Videotape That Will Not Play Back, or Resurrection Bus (1980)

Ralph Hocking

Due to manufacturing errors, changes in videotape machinery and videotape formats, and not a hell of a lot of experience on anyone's part, there are some immediate, obvious problems that need solving when considering the transfer of early videotapes.

The tapes concerned are the reel-to-reel tapes made from the late 1960s to about 1975. This includes several formats of ½", ¼", and maybe some 1" videotape. Information on 2" tape needs only to be transferred by commercial services still currently available.

The tapes were the result of the invention of low-cost video production capability, and made by people who were concerned with the issues of their time. Resurrecting must be based on the tape-makers' views of the importance of the original intention, and not on current commercial or exhibition value, or collectability. Most, if not all, of these tapes can be identified in quick fashion. It is imperative to find the original tapes. That seems to be possible, also.

The focus should be on the tape-makers and whether or not the makers want the material to be salvaged. Collectors, including museums, galleries, media centers and individuals, should be resources for identifying the tapes and not primary sources of tapes for salvage in any preservation effort.

The makers should be actively involved in this preservation attempt as salvagers, financiers, or in any way they can. Otherwise our efforts are not reasonable.

The original material needs to be identified. The makers need to decide how and whether they want their work to be revived. We need to offer the possibility of help to that end.

Once the material is identified and the cooperation of the maker or current owner is obtained, then an attempt could be made to redo the old stuff as new stuff, keeping the original video intention intact.

A vehicle would be outfitted with all early formats. I think I have them. My machines need to be reconditioned, but I think I also have most parts needed. To facilitate transfer, frame synchronizers, ProcAmps, magic, candles and prayer rugs would be available. The transfers would be made to a current videotape format. The choice of what format to transfer to would be up to the owner of the material. All current videotape formats would be available.

The vehicle would travel to the tapes, cancelling the potential for loss of the original material. In the case of cities – i.e., New York – a place would be found to house the service for the time needed for the transfers.

To facilitate the transfers, it seems desirable to send ahead a tape-cleaning machine. The machine would be constructed so that the owners of the tapes could understand and use it to clean the tapes. Cleaning is the most time-consuming part of the process of

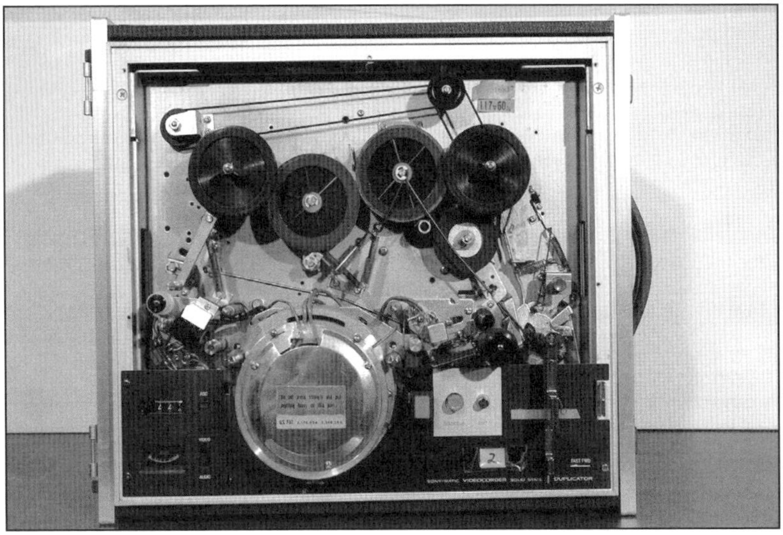

Figure 1. A Sony CV 2200 ½" open-reel deck with the top cover removed.

retrieving the information, and so could be the most expensive part of the process. The videotape makers/owners or their designees would be expected to be involved with the process. Perhaps regional media centers could offer this service at a low cost for owners who couldn't serve themselves. We would still supply the means of cleaning.

This would be a one-shot deal, not an ongoing service. Once the tapes are transferred to a current format there is no further need for the service.

This service is offered in the spirit of the original intention of the tape-makers and of the times when the tapes were made. It is, in fact, an attempt to reclaim the past lost to the ambition of makers of machines and consumer hunger for new technology. The past was lost not only with hardware development, but also with faulty videotape. I don't blame the originators of the machines or gooey videotape manufacturers. None of us knew what we were doing. Do we ever? However, I would expect the makers of both to be spiritually and financially involved in this effort.

This is, of course, a rough proposal, but also a commitment on my part to develop the idea further if there is reason to do so. It seems to me that the project could be limited to this state or has a potential to cover the entire country. Interest and moneys available would determine the scale of involvement.

If the idea is sensible to this group, then I am willing to do the research, find out how much it will cost, and what else it takes to develop a workable definition of the idea.

If not, then I won't.

I can be reached by mail at the TV Center and by phone. I prefer [e]mail.

X

Program Request Sheet **Page 2 of 10**

For each program summarized on page one, submit: A Program Request Sheet, a program budget, and any supplementary materials you wish to attach. If you are requesting funds for more than one program, reproduce this page (either photocopied or typed) for each program.

1. Name of organization Experimental Television Center Ltd.

Program title (as shown on page one) Technical Innovations Projects: Dave Jones

Program priority number (as shown on page one) ______

Name and telephone of person responsible for this program Dave Jones 607 723-9509

Program starting date (as shown on page one) July 1, 1976 Ending date December 31, 1976

Location (facility and address) Experimental Television Center 164 Court St. Binghamton NY

County(ies) in which services will be offered. If more than one, estimate the dollar amount of requested NYSCA funds to be used per county Broome; equipment will be available to artists from throughout the State.

2. Complete description of program or activity within this space.

This proposal is a series of three projects involving modifications to existing video equipment or the design and construction of new systems.

1. Wipe Generator Bank: One of the basic video tools has been the SEG. We would like to take one aspect of this system and expand it to increase image making potentials for the artist. The wipe effect has been largely overlooked because it is capable of nothing more than split screen wipes and corner inserts. With this proposed bank the possibilities are exciting. With nine inputs, nine distinct images can be displayed on one screen simultaneously; combinations of images are innumerable.

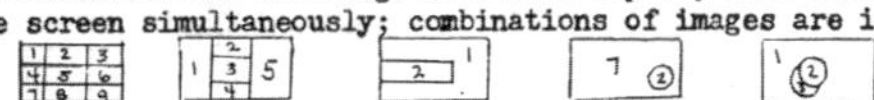

2. Digital Raster Manipulation (video storage unit): The Center has been concentrating efforts in the area of colorizers, keyers and other manipulations of the video signal that generally alter the image in ways which do not control the raster directly. The exception to this is the simple raster control through the application of magnetic fields outside the CRT. We would like to investigate the use of low cost memory and other digital components in a raster manipulation unit. The proposed unit will: 1. still frame and store frame 2. store and recall any portion of the raster in any combination 3. move a section of the raster to any other position 4. reverse images around horizontal and vertical axis 5. enlarge image size to double or reduce to half size. This unit will have low but usable resolution; matrix resolution will vary from 64x64 to 128x128 and in 1/2 screen size 256x256. It will also vary from 16 to 32 shades of gray. The cost of 525x525 resolution is prohibitive because of the enormous storage capacity required. It must be emphasized that this unit is a basic idea which can be added to; eventually it will have capabilities similar to the Rutt/Etra, using concepts not included in the Rutt/Etra design.

3. Modifications to the Jones Colorizer: The development of the Jones was supported by the Council in 1974-75. This proposal includes modifications to the basic design to incorporate digital switching, an internal sync generator and new control panel.

The results of these three projects in the form of equipment and systems will be made available for use by visiting videomakers at the Center.

Attached: vita: Dave Jones
Itemized Budget

3. Attach a detailed program budget following the Budget Instructions in the Guidelines.

Enter from your attached budget: Cost of program $ 10,175.00 Amount requested $ 10,175.00
(These figures should agree with those shown on page one.)

ETC proposal on behalf of Dave Jones to the New York State Council on the Arts in 1976 for a Wipe Generator Bank, the Digital Raster Manipulation Unit and modifications to Jones Colorizer.

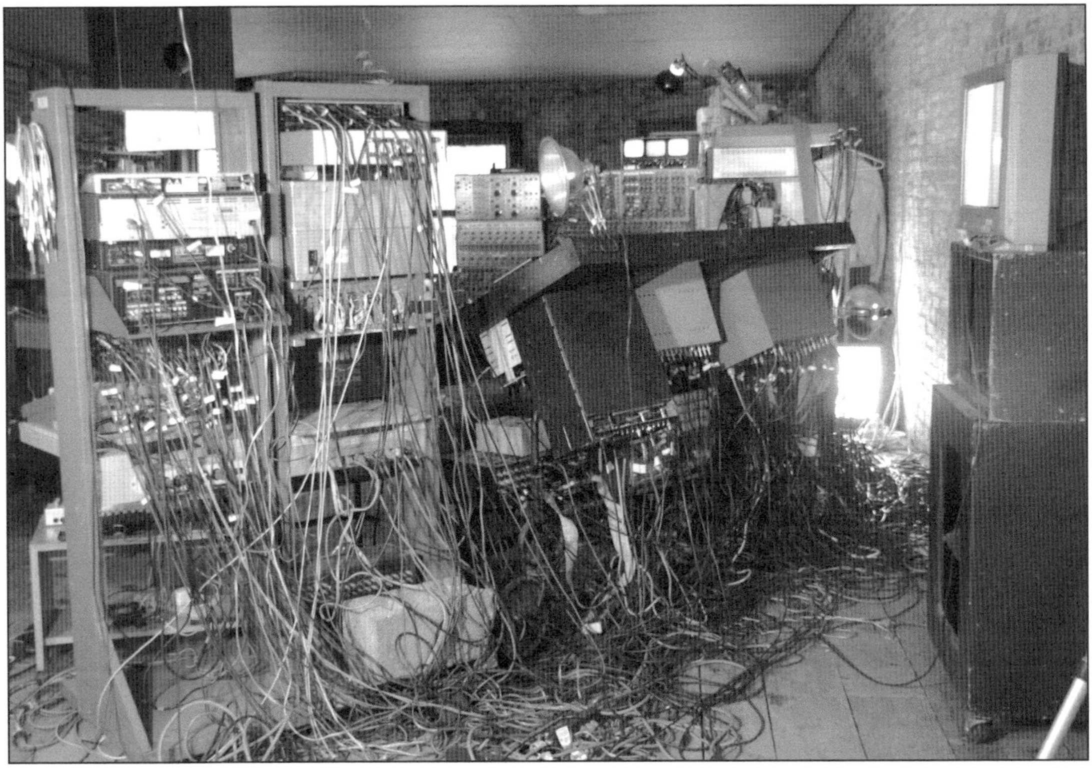

A look behind the Image Processing System at the ETC studio.
(photo. Olivia Robinson. courtesy. Experimental Television Center).
For another view please see Color Plate 43.

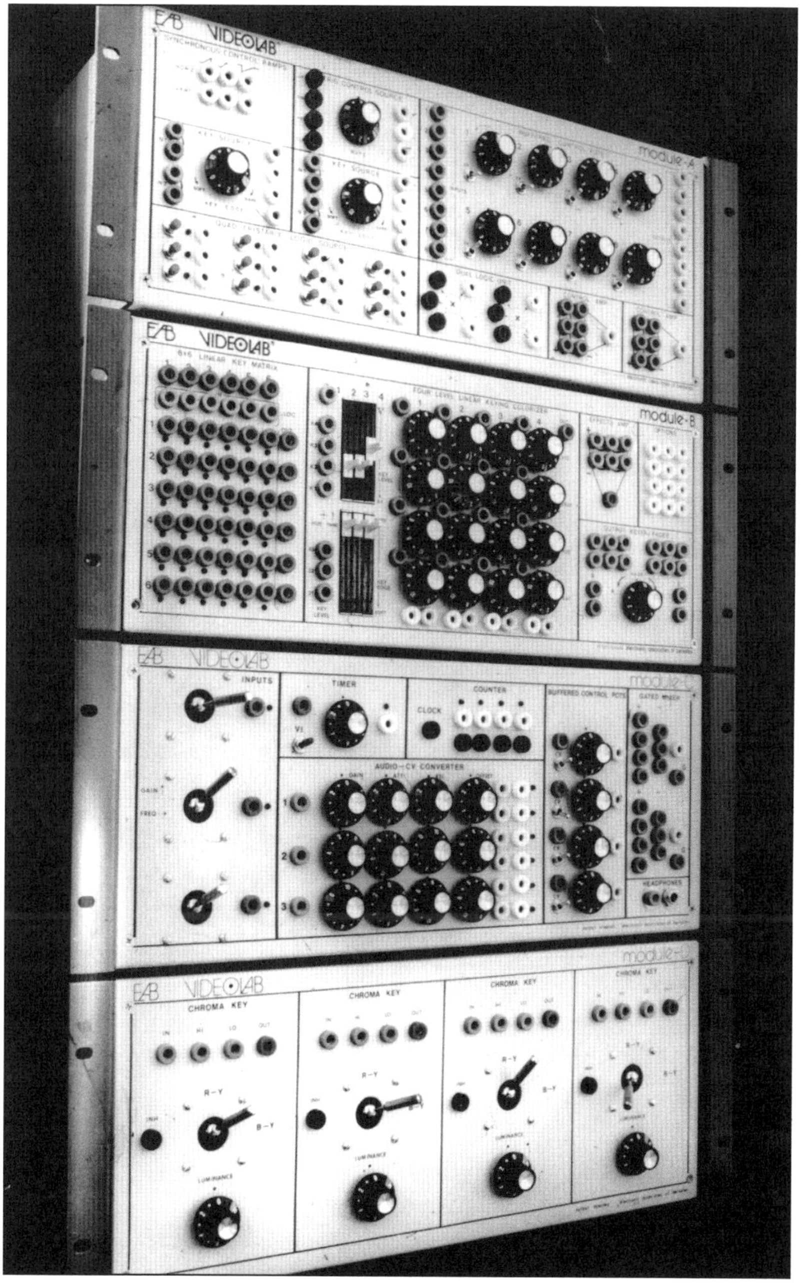

Product literature from Electronic Associates of Berkeley concerning the Hearn Video Lab, designed by Bill Hearn (n.d.). For photos of two of the modules please see Color Plates 33 and 34.

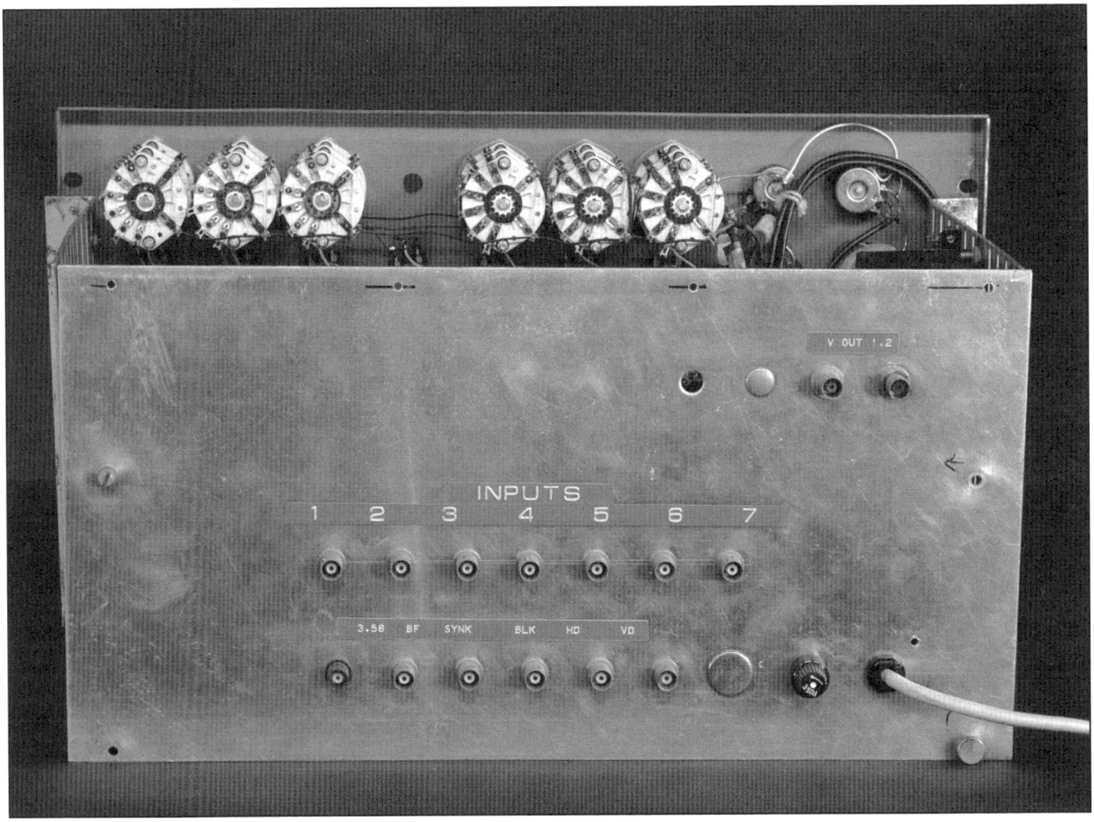

Back panel of Paik/Abe Video Synthesizer. (photo. Olivia Robinson. courtesy. Experimental Television Center).

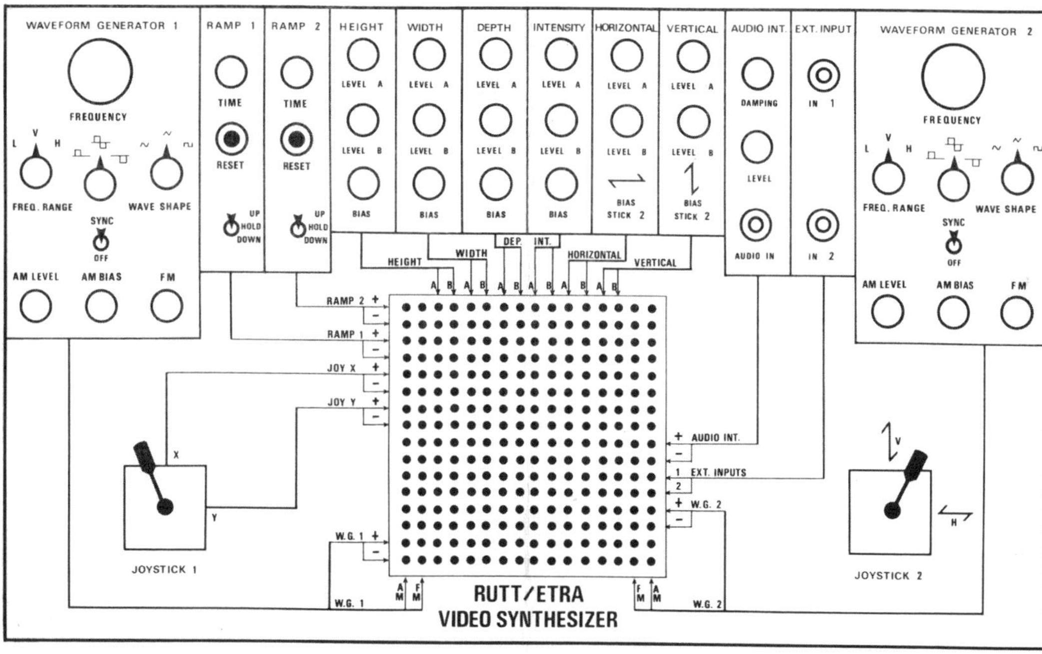

The front panel diagram for Rutt/Etra Video Synthesizer
from product sales literature (n.d.).
Also see Color Plate 31.

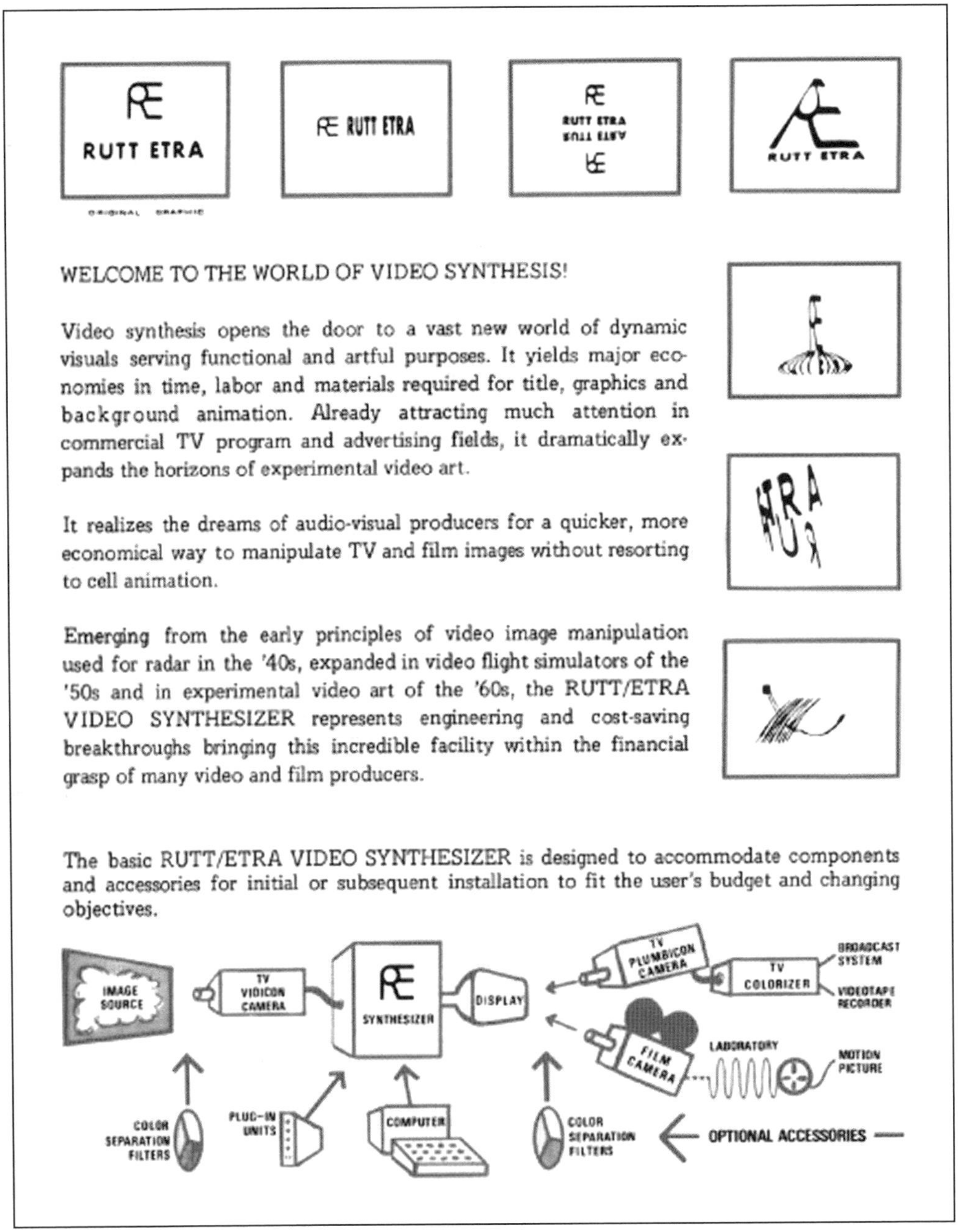

WELCOME TO THE WORLD OF VIDEO SYNTHESIS!

Video synthesis opens the door to a vast new world of dynamic visuals serving functional and artful purposes. It yields major economies in time, labor and materials required for title, graphics and background animation. Already attracting much attention in commercial TV program and advertising fields, it dramatically expands the horizons of experimental video art.

It realizes the dreams of audio-visual producers for a quicker, more economical way to manipulate TV and film images without resorting to cell animation.

Emerging from the early principles of video image manipulation used for radar in the '40s, expanded in video flight simulators of the '50s and in experimental video art of the '60s, the RUTT/ETRA VIDEO SYNTHESIZER represents engineering and cost-saving breakthroughs bringing this incredible facility within the financial grasp of many video and film producers.

The basic RUTT/ETRA VIDEO SYNTHESIZER is designed to accommodate components and accessories for initial or subsequent installation to fit the user's budget and changing objectives.

A description of the Rutt/Etra Video Synthesizer system from product sales literature (n.d.).

United States Patent

Siegel

[15] **3,647,942**

[45] **Mar. 7, 1972**

[54] **VIDEO COLOR SYNTHESIZER**

[72] Inventor: **Eric J. Siegel**, c/o Tom Tadlock 2143 Pine Street, San Francisco, Calif. 94115

[22] Filed: **Apr. 23, 1970**

[21] Appl. No.: **31,090**

[52] **U.S. Cl.** **178/5.4**, 178/6.8

[51] **Int. Cl.** **H04n 9/02**

[58] **Field of Search** .... 178/6.8, 5.2, 5.4, 69.5 CB

[56] **References Cited**

UNITED STATES PATENTS

3,551,589	12/1970	Moskovitz	178/5.4
2,982,814	5/1961	Fine et al.	178/6.8
2,874,217	2/1959	Diehl	178/69.5 CB

Primary Examiner—Richard Murray
Attorney—Hubbell, Cohen & Stiefel

[57] **ABSTRACT**

An apparatus for synthesizing pseudocolor video signals from composite monochromatic video signals. Means are provided for producing a color burst signal on the back porch of the pseudocolor video signal from a monochromatic horizontal sync pulse. Such means includes a color subcarrier oscillator, a pulse generator which stretches the horizontal sync pulse to accommodate the proper period of the subcarrier frequency and slightly delay its occurrence, and a burst gate which produces the burst frequency from the subcarrier and stretched pulses. A pseudochrominance portion of the color video signal is produced by means which phase modulate the color subcarrier with the monochromatic video signal to produce a pseudohue signal and mix this hue signal with the video information portion of the monochromatic signal. An output mixer is also provided for mixing this pseudochrominance signal with the color burst and the monochromatic video signal to provide the pseudocolor video signal. Upon reception by a color video receiver, the monochromatic picture is reproduced in subjective color, the hues therein varying in accordance with the changes in the gray level of the monochromatic video information.

15 Claims, 3 Drawing Figures

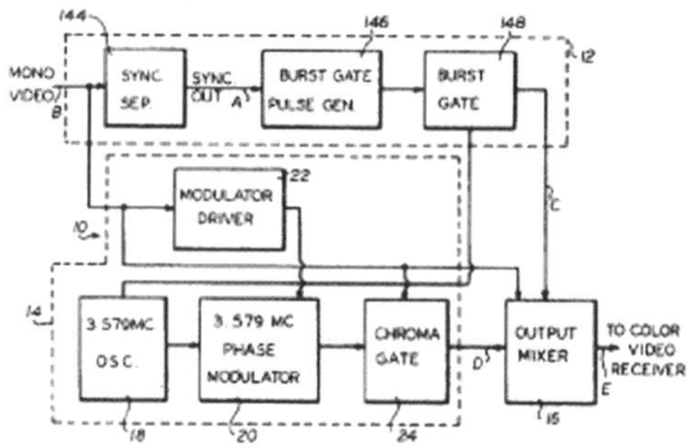

The patent for Eric Siegel's Video Color Synthesizer, filed April 23, 1970.

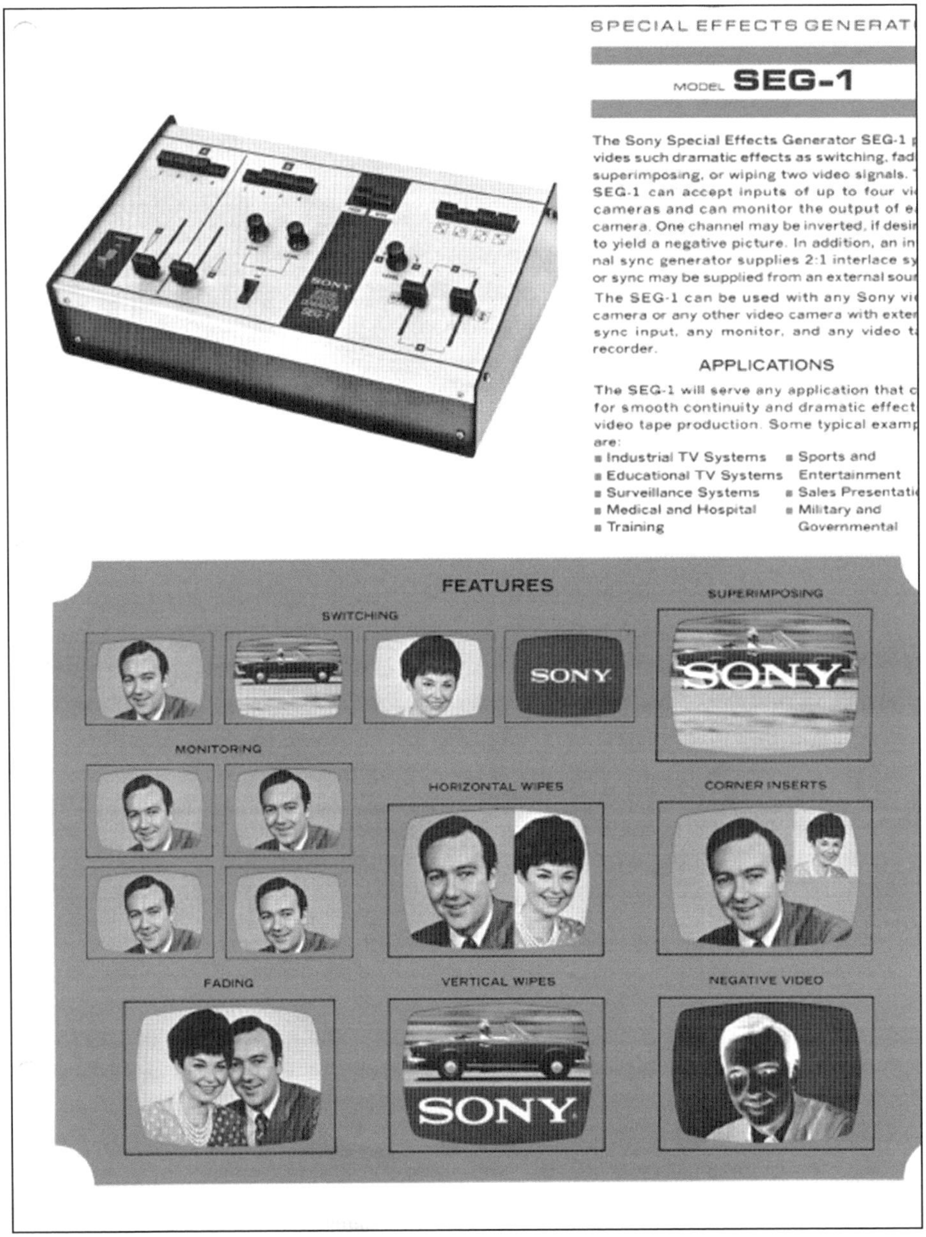

Product literature showing the types of images that could be generated with the Sony Special Effects Generator, SEG-1, available in 1971 (n.d.).

AUTHOR BIOGRAPHIES

RICHARD BREWSTER

Richard Brewster has a degree in Computer Information Systems and has been working since 1993 as a software developer, currently in Charlottesville, Virginia. He has been involved with electronics since high school and in the early 1970s began building modular synthesizers. He worked for several years at the Experimental Television Center on many projects, among them the construction of an audio synthesizer and control tool for the ETC system. Today he builds customized analog synthesizer equipment for himself, and enjoys participating in the Internet community for analog synthesis. He occasionally performs electronic music with his synthesizer. He is the owner of Rabbit Software and offers independent software consulting services focusing on server-side application development.

JON CATES

Jon Cates is the Chair of the Film, Video, New Media & Animation Department at the School of the Art Institute of Chicago. His experimental new media art projects are presented internationally in exhibitions and events from Berlin to Beijing, Cairo to Chicago, Madrid to Mexico City and online. His writings on media art histories also appear online and in print publications, as in recent books from Gestalten, The Penn State University Press and Unsorted Books. He founded the Phil Morton Memorial Research Archive to freely and openly share Morton's original research into the origins of new media and video art.

MICHAEL CENTURY

Michael Century is Professor of New Media and Music in the Arts Department at Rensselaer Polytechnic Institute, joined in 2002. Long associated with The Banff Centre for the Arts, Century founded the Centre's Media Arts Division in 1988. In this position, he was the instigator of the 'Art and Virtual Environments' project (1991–1994). From 1993–1996, Century was a program manager at the Canadian Centre for Information Technology Innovation (CITI). From 1996-98, he served as policy advisor to the federal department of Canadian Heritage, and has been a consultant on new media and cultural policy for various public and university sector clients, including the Rockefeller Foundation.

MICHAEL CONNOR

Michael Connor is a New York–based writer and curator with a focus on artists' responses to cinema and new technologies. Connor previously worked as Curator at FACT, Liverpool, and Head of Exhibitions at BFI Southbank in London. While at the BFI, he developed an interactive moving-image archive designed by Adjaye/Associates as well as a gallery dedicated to artists' film, video, and new media. From 2002 to 2005, Connor worked as a curator at FACT in Liverpool. He is Visiting Curator at Cornerhouse (Manchester).

JEREMY CULLER

Jeremy Culler is an Adjunct Instructor at the State University of New York at Fredonia, where he teaches courses in art history and film studies. He holds a PhD in Art History from Binghamton University. Dr. Culler has recently presented papers at Alfred University, the Frick Collection, the Institute of Fine Arts at New York University and Binghamton University. His research focuses on the discursive field of electronic, time-based media.

LENKA DOLANOVA

Lenka Dolanova is a curator, journalist, and researcher, with interests in art, ecology, and media. She received her PhD at the Film and TV School of Academy of Performing Arts in Prague in 2009. She was Fulbright researcher at the School of the Art Institute in Chicago (2005–2006). Her book about video artists Steina and Woody Vasulka was published in 2011. She works as a curator at Regional Gallery of Vysočina, Czech Republic, and develops projects at the residency center Kravín Rural Arts in Bohemia.

CAROLINA ESPARRAGOZA

Carolina Esparragoza was born in Mexico City in 1977, where she currently lives and works. She studied Visual Arts at the Escuela Nacional de Pintura, Escultura y Grabado-ENPEG La Esmeralda in Mexico. Her work has been exhibited at the Feria Internacional de Arte Contemporáneo ARCO 05 in Madrid, Spain (2005), l'Institut du Mexique in Paris, France (2004), Caja Negra, MUCA-C.U. Universidad Nacional de México (2002), and Museo de la Ciudad de México (2000), among other venues in Mexico and abroad.

JEAN GAGNON

Jean Gagnon is the Director of Collections at the Cinémathèque québécoise. Previously, he served as Executive Director of the Daniel Langlois Foundation for Art, Science, and Technology for ten years. For the Foundation he organized, as curator, the exhibition 'The Body of the Line: Eisenstein's Drawings', presented in Montreal and New York in 1999–2000. He was also the producer of *Digital Snow*, a DVD-ROM on Michael Snow's work. Before joining the Foundation, he was the Associate Curator of Media Arts at the National Gallery of Canada in Ottawa.

KATHY HIGH

Kathy High is an interdisciplinary artist working with science and time-based arts. She produces videos, performances and installations about gender and technology, empathy and animal sentience. She has received awards including fellowships from Guggenheim Memorial Foundation (2010), Rockefeller Foundation, and National Endowment for the Arts, and has had artist residencies with SymbioticA at the University of Western Australia (2009–10), and in Hong Kong with the Asian Arts Council (2005). High is Professor of Video and New Media in the Department of Arts at Rensselaer Polytechnic Institute, Troy, New York – a department specializing in integrated electronic arts practices.

RALPH HOCKING

Ralph Hocking is Professor Emeritus of the Cinema Department at Binghamton University. He has been a leader in the field of video art since 1968, founding what was one of the first campus-based programs. In 1970 he established the independent, nonprofit Experimental Television Center. He has served as consultant, advisor and panelist with the Intercultural Film and Video Fellowship Program of the Rockefeller Foundation, Massachusetts Art and Humanities Foundation, Anthology Film Archives, New York State Council on the Arts, the Society for Photographic Education, the New York Foundation for the Arts and many museums and galleries. His personal creative work has been exhibited internationally, is included in Video Databank's *Surveying the First Decade* anthology and is in the collection of the Museum of Modern Art.

SHERRY MILLER HOCKING

Hocking has served as Assistant Director at the Experimental Television Center since 1972. She has been active in the field of video preservation since the 1980s, serving on the advisory board of Independent Media Arts Preservation Project and of Migrating Media. In 1993 she created and manages the Video History Project, a website featuring documents concerning the early history of video art and community television. Hocking served as project director and cocurator for *ETC: 1969–2009*, a DVD anthology of over 100 artists' works created through the Residency Program at ETC. She is presently collaborating with the Rose Goldsen Archive at Cornell University on the acquisition of the ETC tape and print archives.

MONA JIMENEZ

Mona Jimenez is an artist and educator who has been organizing and advocating for the preservation of media art and community media since the 1980s. She is an Associate Arts Professor and Associate Director in the Moving Image Archiving and Preservation Program at New York University, where she teaches collection management and the preservation of video and digital works, including time-based media art. Recent projects include developing a model for activist archiving of video collections and collaborating with colleagues in Ghana for training in audiovisual archiving and for development of a digital repository of audio and video materials.

DAVE JONES

Dave Jones is a video artist and engineer who has been involved in the video art field since 1971, starting with VideoHeads in Amsterdam, and then the Experimental Television Center several years later. He spent many years designing image-processing tools for video artists and custom equipment for art installations by some of the most famous artists in the field. His tools have been used by artists around the world, and his machines are in use at some of the largest museums worldwide. His latest venture is to manufacture digital versions of his early video image-processing designs.

LOVID

LoVid is an interdisciplinary artist duo composed of Tali Hinkis and Kyle Lapidus. Their work includes live video installations, sculptures, digital prints, patchworks, media projects, performances and video recordings. They combine many opposing elements in the work, contrasting hard electronics with soft patchworks, analog and digital, or handmade and machine-produced objects. This multidirectional approach is also reflected in the content of their work: romantic and aggressive, wireless and wire-full. They are interested in the ways in which the human body and mind observe, process, and respond to both natural and technological environments, and in the preservation of data, signals and memory.

DONALD McARTHUR

Don McArthur received his PhD in Theoretical Physics from the University of Nebraska. He taught physics at the University of Saskatchewan, the State University College at Cortland and Binghamton University. He was associated with the Experimental Television Center beginning in 1974 as a researcher in electronic design. He worked on collaborative projects with Walter Wright, Paul Davis, David Jones and Ralph Hocking of the Center, and with Steina, Woody Vasulka and Jeffrey Schier. He designed the Spatial and Intensity Digitizer, which was exhibited at Ars Electronica in Linz in 1992.

JOHN MINKOWSKY

John Minkowsky is an independent critic and curator of media art. He received undergraduate and graduate degrees from the Department of Media Study at the University of Buffalo. In 1976, he became the Video/Electronic Arts Curator and Music Programmer at Media Study/Buffalo, and his work has been widely published in exhibition catalogs and media arts journals such as *Afterimage* and *FilmQuarterly*. He has received recognition for his writing on media from the New York State Council on the Arts and the National Endowment for the Arts, and has been an Andrew Mellon Fellow in Critical Studies at the California Institute of the Arts. He is writing a history of the role of public television in the development of video art.

TIMOTHY MURRAY

Timothy Murray is Professor of Comparative Literature and English, Director of the Society for the Humanities, and Curator of the Rose Goldsen Archive of New Media Art at Cornell University. He is Managing Co-Moderator of the -empyre- new media listserv, and cocurator of CTHEORY MULTIMEDIA. His books include *Digital Baroque: New Media Art and Cinematic Folds* (Minnesota, 2008), *Zonas de Contacto: el arte en CD-Rom* (Centro de la Imagen, 1999), *Drama Trauma: Specters of Race and Sexuality in Performance, Video, Art* (Routledge, 1997), and *Like a Film: Ideological Fantasy on Screen, Camera, and Canvas* (Routledge, 1993).

MARISA OLSON

Marisa Olson's work combines performance, video, drawing, and installation to address the cultural history of technology, the politics of participation in pop culture and the aesthetics of failure. Her work has been presented by the Whitney Museum of American Art, Centre Georges Pompidou, the British Film Institute and the Sundance Film Festival, among many others. Her critiques of contemporary art and digital visual culture have extended to writing for *Flash Art, Art Review, Afterimage, Planet* and *Art on Paper* to curating exhibitions and programs at the Guggenheim, San Francisco MoMA, White Columns, Artists Space, the Performa Biennial, SF Camerawork and Rhizome. She lives in New York City.

CHRISTIANE PAUL

Christiane Paul is Associate Professor of Media Studies at The New School, New York, and Adjunct Curator of New Media Arts at the Whitney Museum of American Art. Her books include *Context Providers – Conditions of Meaning in Media Arts* (Intellect, 2011), coedited with Margot Lovejoy and Victoria Vesna; *New Media in the White Cube and Beyond* (University of California Press, 2008); *Digital Art* (Thames and Hudson 2003/2008). Among her exhibitions are 'Cory Arcangel: Pro Tools' (May 2011); 'Eduardo Kac: Biotopes, Lagoglyphs and Transgenic Works' (Rio de Janeiro, Brazil, 2010); and 'Feedforward - The Angel of History' (cocurated with Steve Dietz; Laboral Art Center, Gijon, Spain, 2009–2010).

HANK RUDOLPH

Hank Rudolph is a media artist living in Owego, New York. He was Arts Coordinator at the Experimental Television Center from 1984 until 2012, providing individualized instruction to artists participating in the Residency program. For over ten years, he was co-instructor at the International Summer Workshop, a college-credit course sponsored by the Center and the Institute for Electronic Arts (IEA). Rudolph has taught at Binghamton University and the New York State Summer School for Media Arts at Ithaca College. He is presently Technical Specialist at the Department of Cinema at Binghamton University and on the Board of Signal Culture, a new media arts organization.

DAN SANDIN

Daniel J. Sandin is an internationally recognized pioneer of electronic art and visualization. He is director emeritus of the Electronic Visualization Lab and a professor emeritus in the School of Art and Design at the University of Illinois at Chicago. He is continuing his professional activities with Tom DeFanti at Calit2, University of California at San Diego. As an artist, he has exhibited worldwide, and has received grants in support of his work from the Rockefeller Foundation, the Guggenheim Foundation, the National Science Foundation and the National Endowment for the Arts. His video animation *Spiral PTL* is in the inaugural collection of video art at the Museum of Modern Art in New York. In recent years, Sandin has been concentrating on the development of auto stereo VR displays (free viewing, no glasses), and on the creation of network-based tele-collaborative VR art works that involve video camera image materials, rich human interaction and mathematical systems.

TOM SHERMAN

Tom Sherman is an artist and writer working in video, radio and live performance. He founded the Media Arts Section of the Canada Council for the Arts in 1983. His interdisciplinary work has been exhibited and screened internationally, including shows at the Vancouver Art Gallery and the Museum of Modern Art. His most recent book is *Before and After the I-Bomb: An Artist in the Information Environment* (Banff Centre Press, 2002). Currently, Sherman is a professor in the Department of Transmedia at Syracuse University, where he teaches video research, video art history and communication and media theory.

YVONNE SPIELMANN

Yvonne Spielmann is the Dean of Faculty of Fine Arts at Lasalle College of the Arts in Singapore. Her work focuses on media and culture, technology, art, and communication, and in particular on Western/European and nonWestern/Southeast Asian interaction. Her book, *Video, the Reflexive Medium* (MIT Press 2008, Japanese edition by Sangen-sha Press, 2011) was awarded the 2009 Lewis Mumford Award for Outstanding Scholarship in the Ecology of Technics. Her most recent book, *Hybrid Culture*, was published in German by Suhrkamp Press, 2010, followed by the English edition by MIT Press in 2012. Spielmann's work has been translated into French, Polish, Croatian, Swedish, Japanese and Korean. She holds the 2011 Swedish Prize for Swedish-German scientific cooperation.

CAROLYN TENNANT

Carolyn Tennant is a media artist and curator based in Buffalo. Her interdisciplinary practice reflects interests in visual studies and the techno-cultural history of art. In 2006 she became Media Arts Director at Hallwalls Contemporary Arts Center, where she is responsible for overseeing programs including archival initiatives, digitization of analog records, and video preservation projects. In 2007 she helped coordinate Migrating Media, a statewide partnership project protecting our analog collections and audiovisual heritage. She has presented at international conferences and national venues on these projects as well as her research on the regional histories of electronic arts in upstate and western New York.

JACK TOOLIN

Jack Toolin is a transdisciplinary artist who works both independently and collaboratively. He teaches art history, media studies and photography at Pratt Institute and Pace University. He has lectured widely, at institutions such as the Rhode Island School of Design; the University of California at Berkeley; the San Francisco Art Institute; Emerson College, Boston; Kibla Multimedia Center, Slovenia; and the Museum of Contemporary Art Rijeka, Croatia. He was a founding member of the new media collective C5 (1997–2007), whose work has been exhibited nationally and internationally.

STEINA VASULKA

Steina Vasulka was born in Iceland and studied at the Music Conservatory in Prague, before coming to the United States in the 1960s. She was one of the cofounders of The Kitchen in New York and played an important role in the emergence of video art. Her art practice investigates video as a medium and takes the forms of tape and multiscreen installations. She has exhibited internationally at such venues as the Whitney Museum of American Art Biennial, Centre Georges Pompidou and at ZKM / Center for Art and Media in Austria.

WOODY VASULKA

Woody Vasulka was born in Czechoslovakia, where he studied engineering and film. Together with Steina, he produced a pioneering body of early video art works. His individual practice involves a precise investigation into the language and codes of the electronic image. His later works explore the meanings of history and memory, presented through processes of deconstruction of the video signal. His work has been exhibited around the world. Vasulka has taught at many institutions, including the University of Buffalo, and is the recipient of numerous awards for his art. Vasulka was awarded an honorary doctorate at the Academy of Performing Arts in Prague in 2011.

HOWARD WEINBERG

Howard Weinberg is a script-doctor, winning independent documentary filmmaker, journalist and educator. He has taught at Columbia University's Graduate School of Journalism and New York University. His PBS credits include *Sports for Sale, First Things First, net.LEARNING, "Topless Cellist" Charlotte Moorman* and *Sid at 90.* A Founding Producer of *The MacNeil/Lehrer Report*, he was Executive Producer of *Listening to America* with Bill Moyers and a producer-writer for CBS *Sunday Morning* and *Sixty*

Minutes. Volunteer president-emeritus and board member of The New York Film/ Video Council, his current project is *TV LAB: License to Create* – a documentary about an innovative era of public television from 1972–1984.

WALTER WRIGHT

Walter Wright is a multimedia artist whose practice includes music, video and dance improvisation. In the early 1970s, Wright worked as a video animator at Computer Image Corporation, collaborating with Ed Emschwiller, whose *Scapemates* won an Emmy in 1971. In 1973–1976, as artist-in-residence at the Experimental Television Center, he pioneered video performance, touring colleges and galleries with the Paik/ Abe Video Synthesizer. Walter develops software and hardware for artists, such as the Video Shredder, a desktop video processor for the TARGA2000. Currently, he is working with Max/MSP, softVNS, and VDMX. He is a cofounder in 1992 of 119 Gallery, the first digital art gallery on the World Wide Web.

INDEX

Note: Figure page references appear in **bold**.

T

Color Plate 23. Video still from *Bathing* by Gary Hill (1977). (courtesy. Gary Hill).

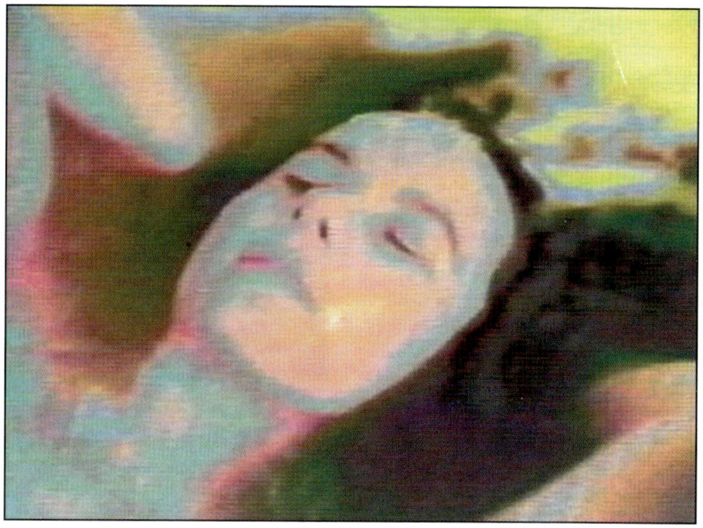

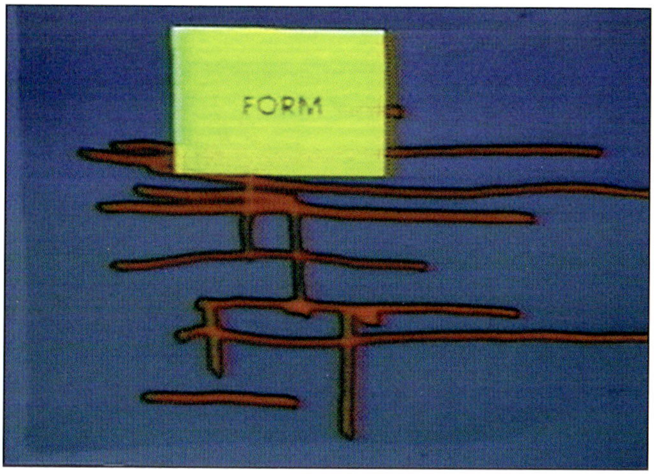

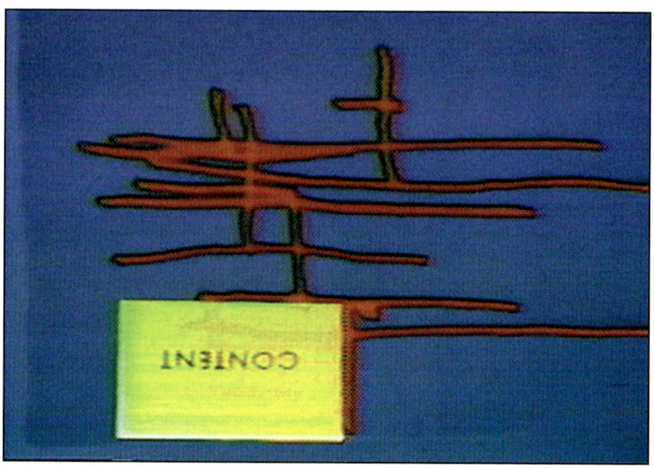

Color Plates 24 and 25. Video stills from *Picture Story* by Gary Hill (1979). (courtesy. Gary Hill).

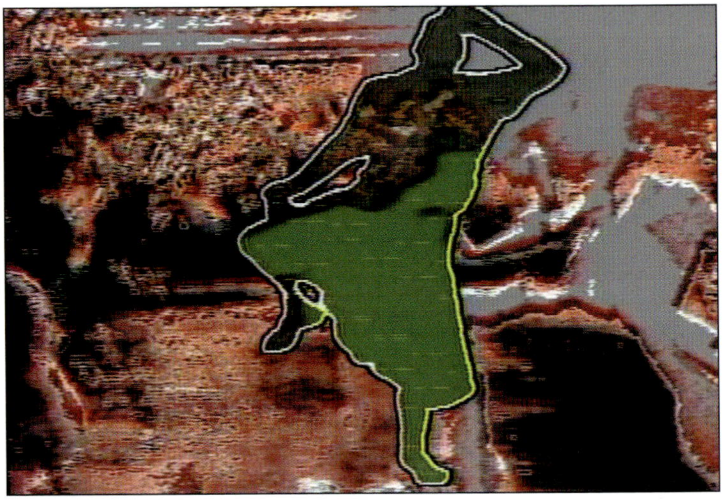

Color Plate 26. Video still from *DA.DA.dance* by Connie Coleman and Alan Powell (1990) which demonstrates use of the Jones Buffer and Colorizer. Music by BeauSoleil. (courtesy. Alan Powell).

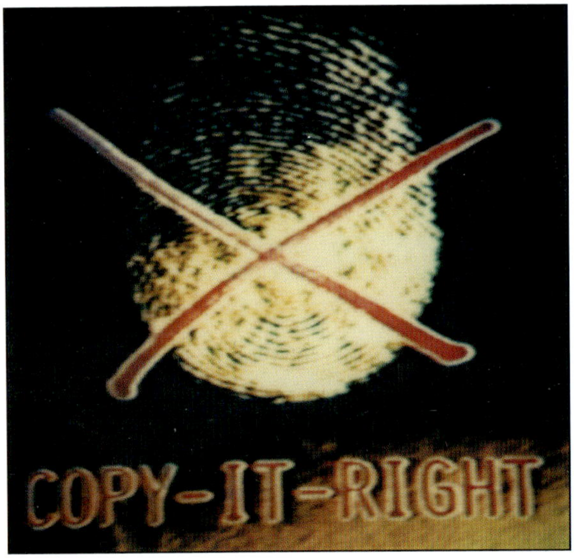

Color Plate 27. Phil Morton's symbol for COPY-IT-RIGHT. (courtesy. Phil Morton Memorial Research Archive).

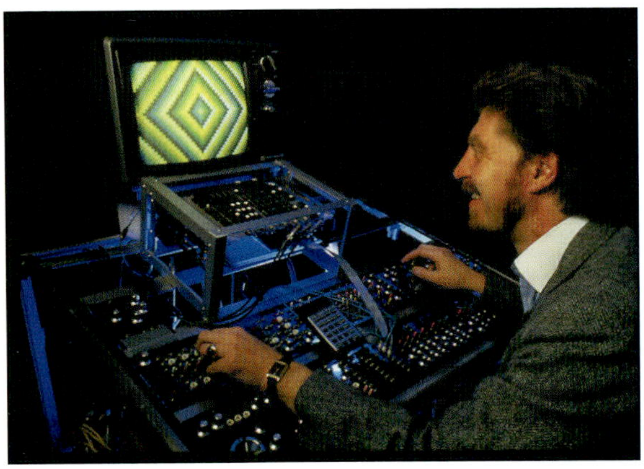

Color Plate 28. Stephen Beck with his Video Weaver instrument (1976). © Stephen Beck, 1976. All rights reserved. www.stevebeck.net. (photo. Steve Beck).

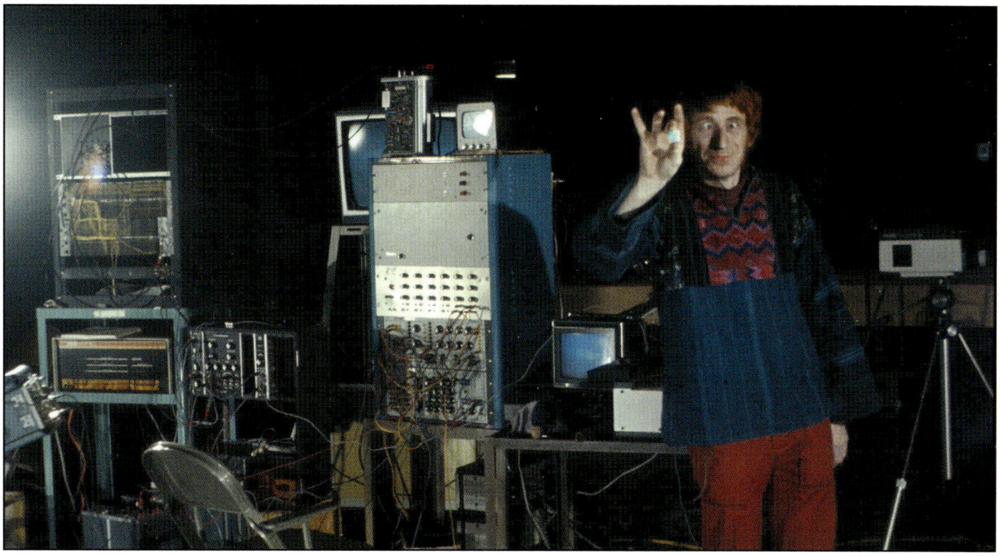

Color Plate 29. As the character Zierot le Fou in his film *Outta Space*, Tom DeWitt manipulates a bright blue dot that became a telescope animated by the Rutt/Etra Video Synthesizer, a component of the Pantomation system (1976–77). (photo. Vibeke Sorensen; courtesy. Tom Ditto).

Color Plate 30. (Below) Inside a Paik/Abe Video Synthesizer constructed around 1973/74. (photo. Olivia Robinson. courtesy. Experimental Television Center).
Color Plate 31. (Right) Product literature describing the functionality of the Rutt/Etra Video Synthesizer System (n.d.)

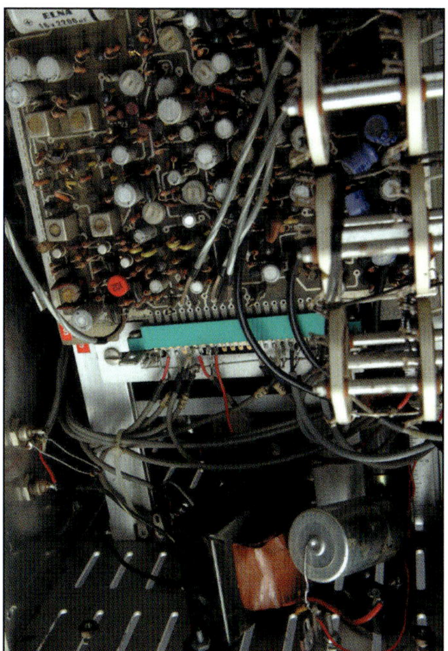

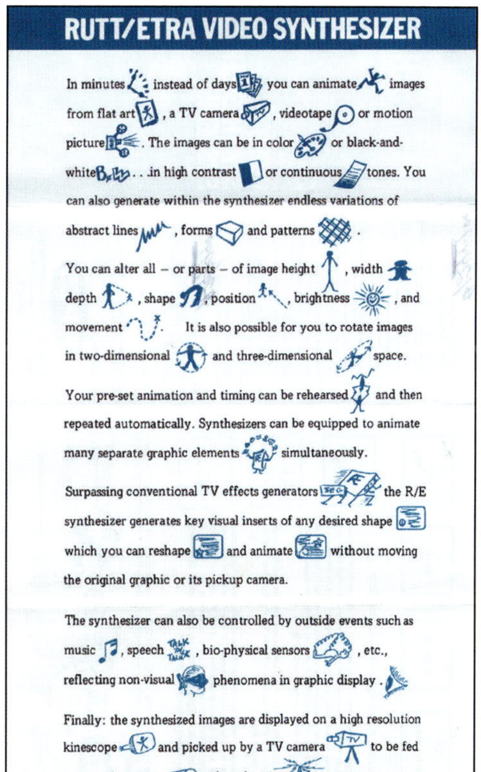

RUTT/ETRA VIDEO SYNTHESIZER

In minutes instead of days you can animate images from flat art, a TV camera, videotape or motion picture. The images can be in color or black-and-white. . .in high contrast or continuous tones. You can also generate within the synthesizer endless variations of abstract lines, forms and patterns.

You can alter all – or parts – of image height, width, depth, shape, position, brightness, and movement. It is also possible for you to rotate images in two-dimensional and three-dimensional space.

Your pre-set animation and timing can be rehearsed and then repeated automatically. Synthesizers can be equipped to animate many separate graphic elements simultaneously.

Surpassing conventional TV effects generators the R/E synthesizer generates key visual inserts of any desired shape which you can reshape and animate without moving the original graphic or its pickup camera.

The synthesizer can also be controlled by outside events such as music, speech, bio-physical sensors, etc., reflecting non-visual phenomena in graphic display.

Finally: the synthesized images are displayed on a high resolution kinescope and picked up by a TV camera to be fed into a videotaping or broadcasting system.

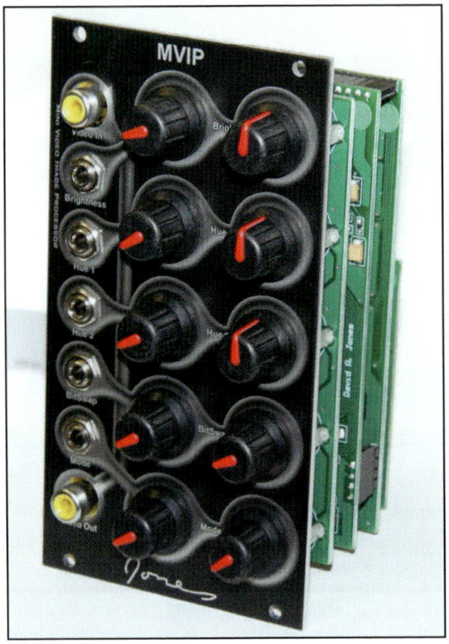

Color Plate 32. The MVIP (Mini Video Image Processor), designed by Dave Jones, is less than three inches wide (2011). (photo & courtesy. Dave Jones Design).

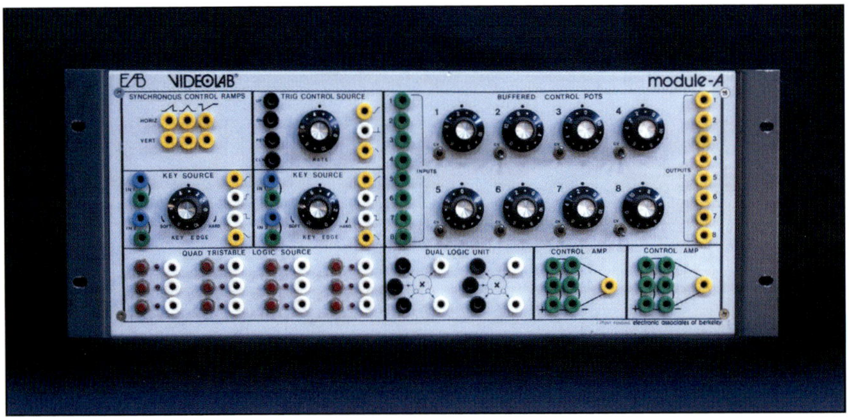

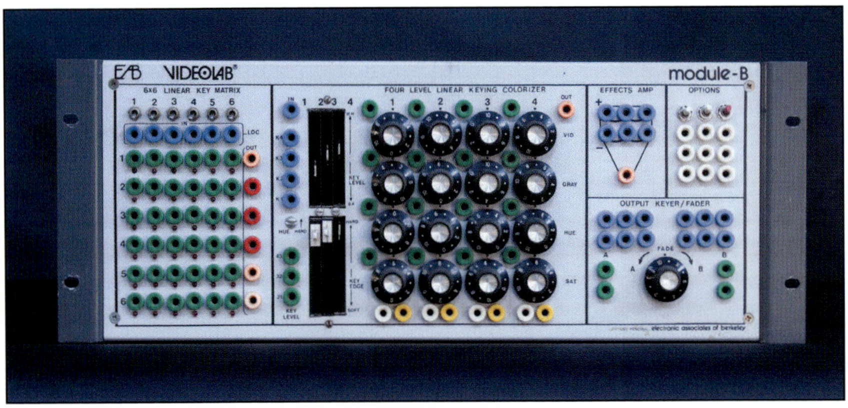

Color Plates 33 and 34. Modules A and B of the Videolab, designed by William Hearn and the Electronic Associates of Berkeley.

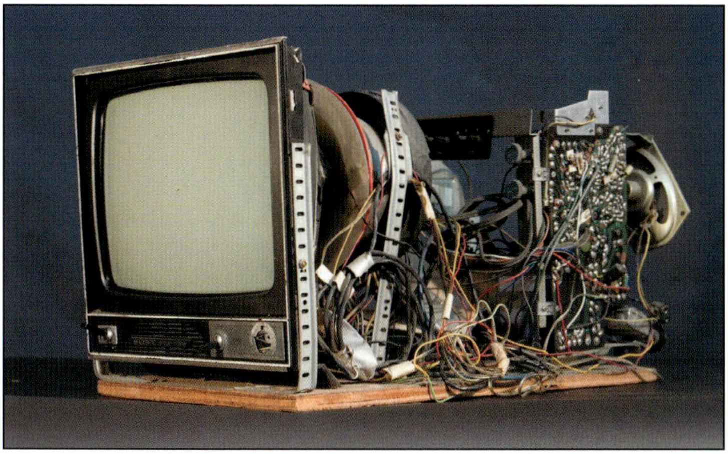

Color Plates 35 (Left) and 36 (Below). Two versions of a Raster Manipulation Unit (aka Wobbulator) in the studio of the Experimental Television Center, dating from the 1970s. (photo. Olivia Robinson. courtesy. Experimental Television Center).

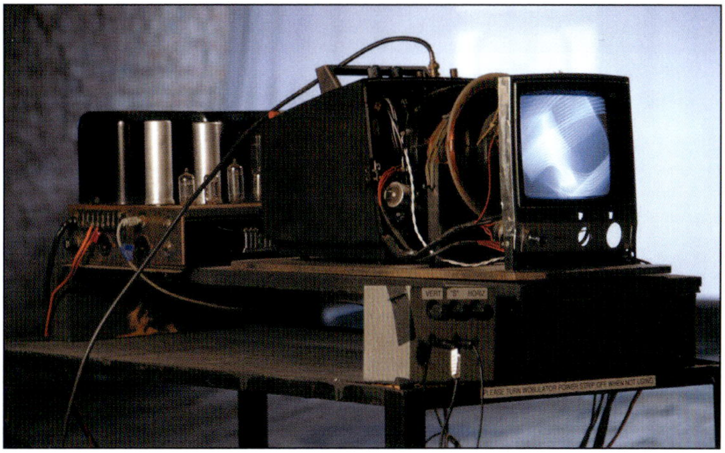

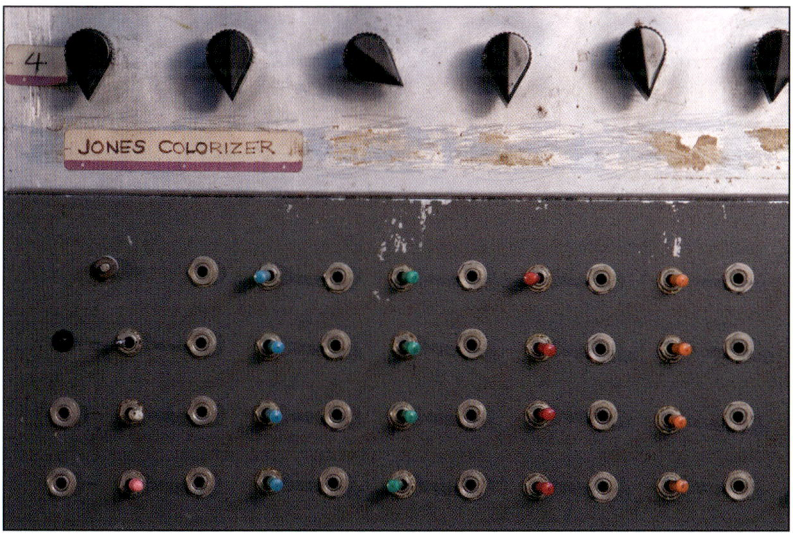

Color Plate 37. Detail of the first Jones Colorizer by Dave Jones, built at the Experimental Television Center. Control panel above, main unit with control voltage inputs below (c. 1974/75). (photo. Olivia Robinson. courtesy. Experimental Television Center).

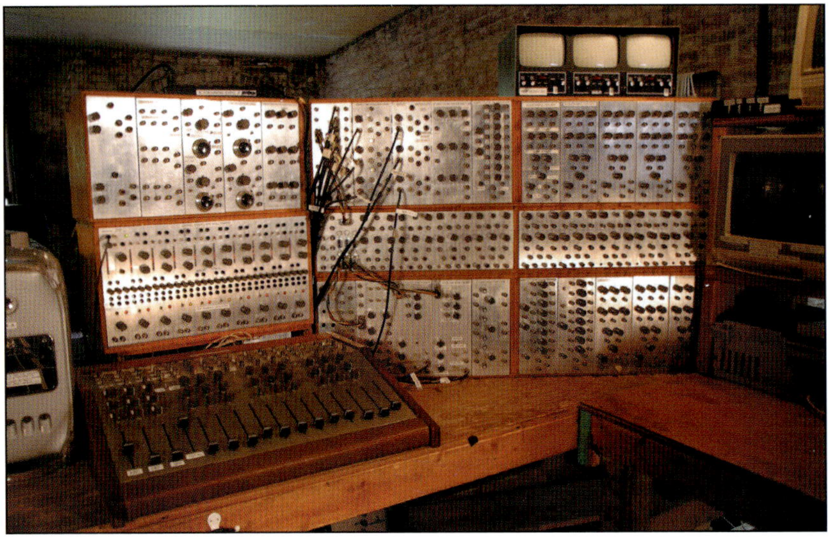

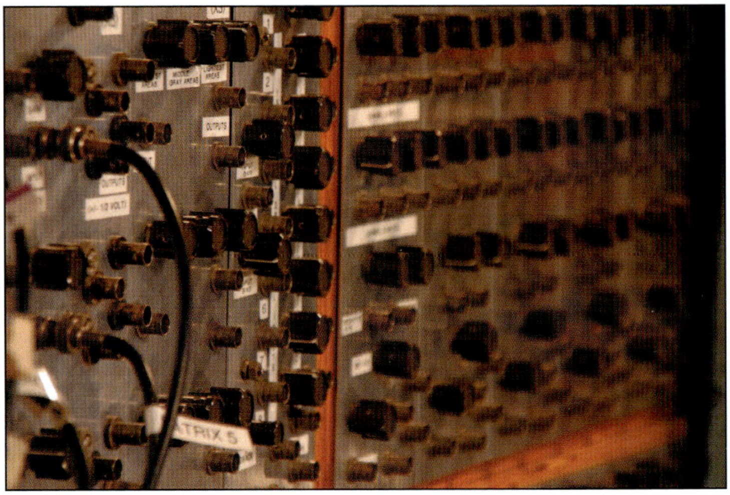

Color Plates 38 (Left) and 39 (Below). Two views of a Sandin Image Processor, built by Dick Sippel, at Experimental Television Center studio. (photo. Olivia Robinson. courtesy. Experimental Television Center).

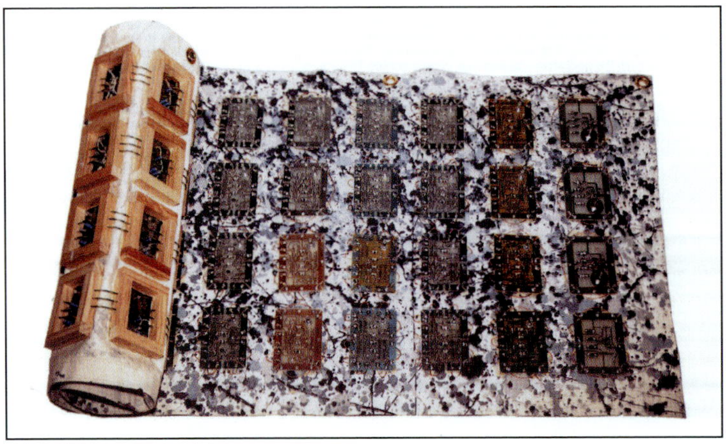

Color Plate 40. *the rollable synthesizer* by Peter Blasser (2002). Image from Peter Blasser documentary file, Collection Foundation Daniel Langlois de la Cinémathèque québécoise. (courtesy. Cinémathèque québécoise).

Color Plate 41. The patch panel for the analog control boxes in the Experimental Television Center studio. Documentation by Richard Brewster (1980).

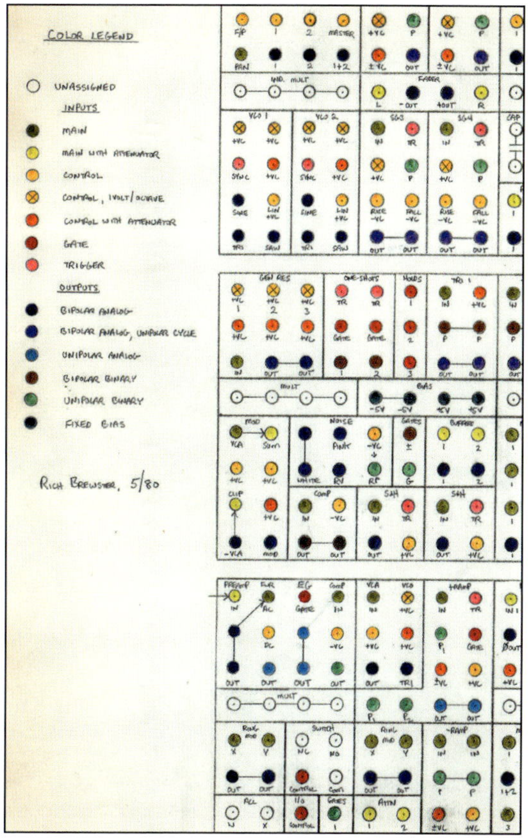

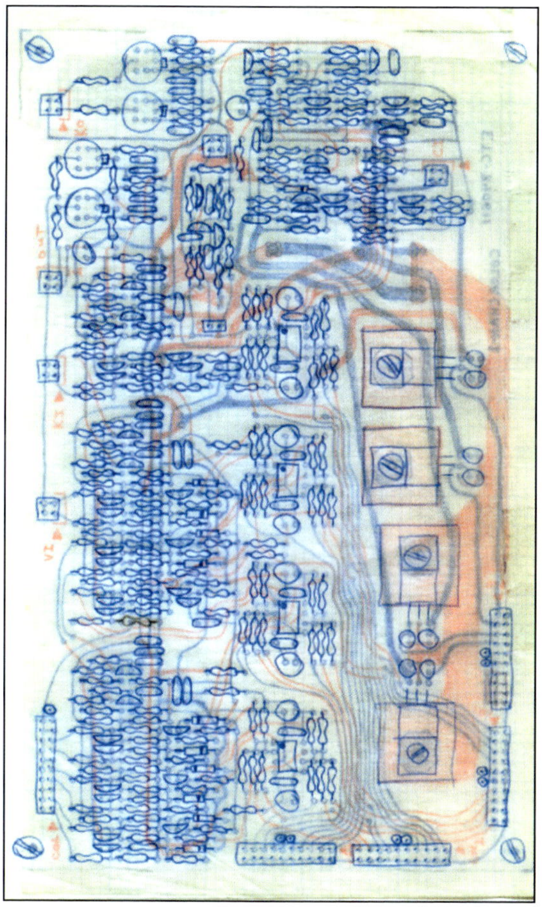

Color Plate 42. Circuit board layout drawing for one channel of the Jones Colorizer designed by Dave Jones for the Experimental Television Center studio (1984).

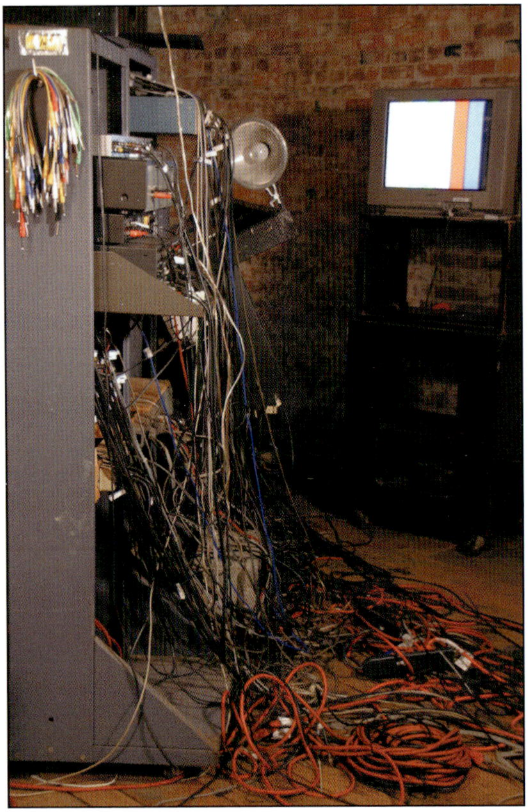

Color Plate 43. A look at the complex cabling behind the system at the Experimental Television Center studio. (2007). (photo. Olivia Robinson. courtesy. Experimental Television Center).

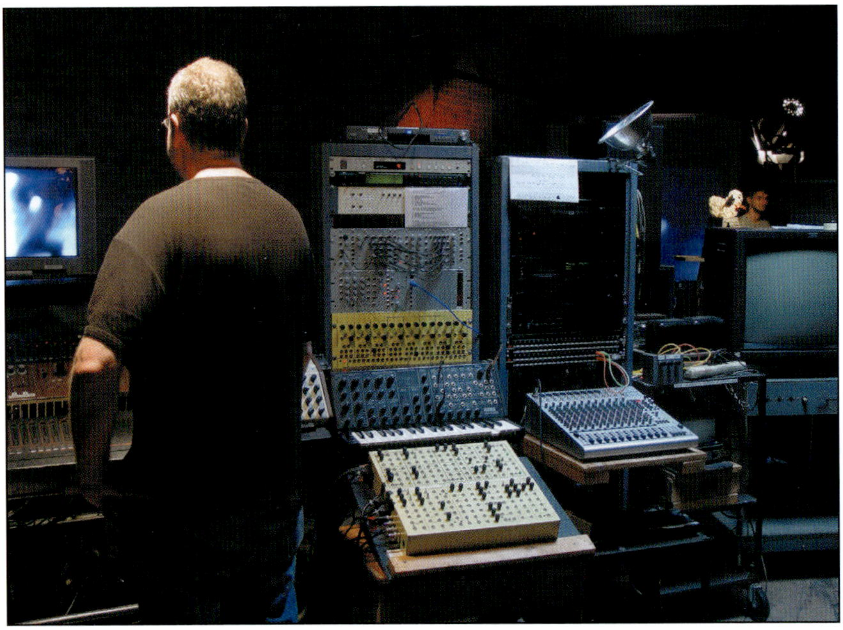

Color Plate 44. Hank Rudolph at the system during the 2005 International Summer Residency at the Experimental Television Center. (photo. Fei Jun & Pamela Susan Hawkins).